DICTIONARY
OF
ENERGY

ENERGY, POWER, AND ENVIRONMENT

A Series of Reference Books and Textbooks

1. Heavy Oil Gasification, *edited by Arnold H. Pelofsky*
2. Energy Systems: An Analysis for Engineers and Policy Makers, *edited by James E. Bailey*
3. Nuclear Power, *by James J. Duderstadt*
4. Industrial Energy Conservation, *by Melvin H. Chiogioji*
5. Biomass: Applications, Technology, and Production, *by Nicholas P. Cheremisinoff, Paul N. Cheremisinoff, and Fred Ellerbusch*
6. Solar Energy Technology Handbook: Part A, Engineering Fundamentals; Part B, Applications, Systems Design, and Economics, *edited by William C. Dickenson and Paul N. Cheremisinoff*
7. Solar Voltaic Cells, *by W. D. Johnston, Jr.*
8. Energy Management: Theory and Practice, *by Harold W. Henry, Frederick W. Symonds, Robert A. Bohm, John H. Gibbons, John R. Moore, and William T. Snyder*
9. Utilization of Reject Heat, *edited by Mitchell Olszewski*
10. Waste Energy Utilization Technology, *by Yen-Hsiung Kiang*
11. Coal Handbook, *edited by Robert A. Meyers*
12. Energy Systems in the United States, *by Asad T. Amr, Jack Golden, Robert P. Ouellette, and Paul N. Cheremisinoff*

13. Atmospheric Effects of Waste Heat Discharges, *by Chandrakant M. Bhumralkar and Jill Williams*

14. Energy Conservation in Commercial and Residential Buildings, *by Melvin H. Chiogioji and Eleanor N. Oura*

15. Physical Cleaning of Coal: Present and Developing Methods, *edited by Y. A. Liu*

16. Flow Phenomena in Porous Media: Fundamentals and Applications in Petroleum, Water, and Food Production, *by Robert A. Greenkorn*

17. Basic Mathematics and Computer Techniques for Coal Preparation and Mining, *edited by Kenneth K. Humphreys and Joseph W. Leonard*

18. Dictionary of Energy, *by Carl W. Hall and George W. Hinman*

Additional Volumes in Preparation

DICTIONARY OF ENERGY

Carl W. Hall
National Science Foundation
Washington, D.C.

George W. Hinman
New Mexico Energy Research and
Development Institute
Sante Fe, New Mexico

MARCEL DEKKER, INC. New York and Basel

Library of Congress Cataloging in Publication Data

Hall, Carl W.
Dictionary of energy.

(Energy, power, and environment ; 18)
1. Power resources--Dictionaries. 2. Power (Mechanics)--Dictionaries. I. Hinman, George W. II. Title. III. Series.
TJ163.8.H34 1983 621.042'0321 83-1956
ISBN 0-8247-1793-7

MARCEL DEKKER, INC.
270 Madison Avenue, New York, New York 10016

Current printing (last digit):
10 9 8 7 6 5 4 3 2 1

PRINTED IN THE UNITED STATES OF AMERICA

PREFACE

Coordinated efforts to resolve the nation's energy problems have brought together workers with diverse backgrounds and interests. Under these circumstances the specialized energy vocabularies from different disciplines can lead to problems in communications. We believe that this dictionary will help resolve these problems. A broad range of words from various sources has been selected with the definitions ranging from the simple to the more complex. Information beyond the simple definition often implied by the word itself is given but an encyclopedic treatment is not presented. We hope that the specialist will understand the necessity to simplify some definitions and also that the reader will have sufficient explanation to facilitate the use of the terms.

The authors gratefully acknowledge the contributions of those who were part of the many publications, presentations, and discussions that preceded this volume. Several organizations with which the authors have worked—education, government, and industry—provided a combined experience of 60 years in the energy field to help build a base for the dictionary. Thanks to Marcel Dekker, Inc., for encouraging the project and for supplying numerous books on energy that were used to compile and define many of the words in this dictionary. We thank our typist, Joanna Farnsworth, and our draftsperson, Karen Parvin, who were diligent workers and, along with the publisher's editorial staff, helped put our drafts in final form.

Carl W. Hall, P.E.

George W. Hinman

DICTIONARY
OF
ENERGY

A

AAAS American Association for the Advancement of Science.

AAES American Association of Engineering Societies.

AAPG American Association of Petroleum Geologists.

AAQS Ambient air quality standards.

ABIOGENIC A substance produced from other than biological means such as abiogenic methane. Also see *Biogenic*.

ABS American Bureau of Shipping.

ABSOLUTE ALCOHOL Usually applied to ethanol (C_2H_5OH) which has 1% or less water, thus absolute is nearly 100% alcohol, or water-free. A 100% alcohol is known as 200 proof. Using normal distillation methods a maximum purity of 95-96% is obtained, and is anhydrous alcohol. Hydrous alcohol has more than 5% water. Also see *Alcohol; Azeotrope; Ethanol; Gasohol; Methanol; Neat Alcohol*.

ABSOLUTE DENSITY The mass per unit volume of a substance which has dimensions of M/L^3, and for which the value is for the actual substance involved and does not include the space between particles. Also see *Apparent Density; Apparent Specific Gravity; Bulk Density*.

ABSOLUTE TEMPERATURE A scale of temperature based on a zero value at absolute zero. Absolute temperature values are used for calculations involving multiplication or division involving temperature (T) or temperature ratios. The common absolute temperature scales are the Kelvin scale (°K = °C + 273) and the Rankine scale (°R = °F + 460). Also see *Celsius; Rankine*.

ABSOLUTE UNITS Various absolute systems of physical units are generally used including primarily the British absolute system, the metric absolute system, and system international units (SI). The absolute systems of units are based on fundamental units of length, mass and time; for the British absolute system, foot-pound-second (fps); for the metric absolute system, centimeter-gram-second (cgs); and for the SI units, meter-kilogram-second. Variations from these are also used.

ABSOLUTE ZERO The temperature at which every system is in its lowest energy state. Neglecting quantum mechanical effects it can be considered to be the temperature at which thermal motion ceases, at 0°K or -273.1°C or -459.6°R.

ABSORBED DOSE The energy absorbed from a flux of radiation by a unit mass of tissue exposed to the radiation. The usual measure of absorbed dose is the rad, which is an absorbed dose of 100 ergs per gram of tissue.

ABSORBENT A substance that absorbs or is capable of absorbing a liquid or a gas, usually through physical, chemical, or molecular means, such as $CaCl_2$ or LiCl. The absorbate penetrates the absorbent with a possible resultant loss of identity of the original material.

ABSORBER (nuclear) A material or device that strongly absorbs nuclear radiation moving through it. The designation must be used in the context of a specific type and energy of nuclear radiation. For example, boron is a good thermal neutron absorber but a poor absorber of gamma rays.

ABSORPTION A unit operation in which a soluble component of a mixture of gases is dissolved in a solid or liquid. The process of the taking up of one substance by another substance, generally by solution. Also see *Absorbent; Adsorption; Sorption*.

ABSORPTION REFRIGERATION A method of providing a cooling effect at the evaporator of a refrigeration system. The system consists of a generator, where heat is added; ammonia is removed from the absorbent ammonia-water solution; a condenser where heat is removed and the ammonia changed to a liquid; and the evaporator, where cooling results with the evaporation of the ammonia (Figure A-1). The unit is often referred to as a gas refrigerator because gas is used to provide the external heat to drive the system. Other absorbents are water-lithium chloride and water-lithium bromide. Absorption refrigeration systems are used for solar energy applications.

ABSORPTIVITY A measure of the portion of radiant energy falling on a body which is absorbed and converted to heat, with the remainder of the energy being reflected or transmitted. The absorptivity depends on the wavelength of the energy and the roughness and color of the absorbing surface.

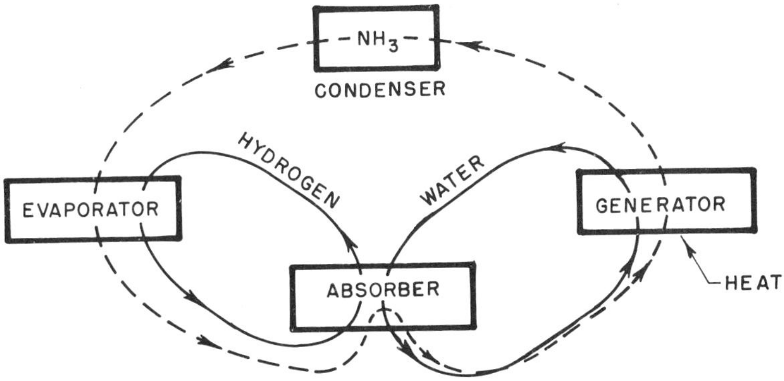

Fig. A-1 Absorption system cycle.

Absorptivity of materials:

Copper	0.26
Iron, 38°C (100°F)	0.45
Stainless steel, 18-8	0.51
Aluminum, polished	0.26
Asphalt pavement	0.85

ABUNDANCE RATIO As applied to nuclear materials, the ratio of the number of atoms of a particular isotope to the total number of atoms.

ACCELERATOR (nuclear) Equipment that is built to accelerate particles to energies that will permit them to initiate nuclear reactions.

ACCEPTOR A substance such as calcined carbonate which absorbs carbon dioxide given off during gasification; heat is liberated in the process. In electronics a semi-conducting substance which consists partially of atoms that bind electrons which would otherwise be free to move through the semi-conducting material.

ACCRETION ENERGY Energy released when matter comes together, such as possibly with the formation of the earth, trapping energy internally. Also see *Gravitational Energy*.

ACCUMULATED DOSE The total dose of radiation accumulated by a person or location over some specified period of time.

ACCURACY The correctness of a measurement, calibration, or a value. Certain commercial practices or standards may be followed and various statistical techniques may be used to represent the accuracy of a numerical value of an instrument. Precision refers to the reproducibility of data by an instrument or procedure.

ACES Annual cycle energy system.

ACETYLENE A colorless, flammable poisonous gas, C_2H_2, which is the first of a series of unsaturated hydrocarbons in which the carbon atom has a triple bond. Acetylene can be made from calcium carbide and water. Acetylene is used as a high temperature fuel source, which burns at a temperature of 3300°C (5972°F).

ACID A chemical compound or material containing hydrogen which will ionize and a part or all of the hydrogen ions may be exchanged for metal or base ions. Acids have a relatively high affinity for water molecules and thus can be considered as a moisture absorber for some applications. The strength of an acid is represented by the concentration of the hydrogen ions, a neutral solution having a pH of 7, with a lower number representing a stronger acid. Also see *pH*.

ACID DRAINAGE The liquid drainage from coal mines or pits resulting from water flow through the mines and mine tailings, in which, for example, the exposed pyrite (FeS_2) breaks down to form an acid (H_2SO_4).

ACID GAS Natural gas which contains hydrogen sulfide, (H_2S), also known as sour gas. Hydrogen sulfide is removed by washing with NaOH alkali solution, or if large amounts absorbed in amine, or as a source of sulfur, if removed. H_2S is an undesirable component if the gas is used as a fuel because it may result in corrosion of equipment and pollution of air, later resulting in acid rain. Also see *Acid Rain; Sweet Gas*.

ACID RAIN Precipitation which has a pH lower than 5.6, which can occur as a result of atmospheric pollution. Atmospheric pollution can occur as a result of combustion. The principal acid anions are sulfate (SO_4^{--}), nitrate (NO_3^-), and chloride (Cl^-). Acid rain can cause inhibited plant and insect growth, deterioration of buildings and equipment, and damage to lakes and streams. Also see *Energy Security Act*.

ACIL American Council of Independent Laboratories.

ACRE A unit of area consisting of 43,560 ft^2, 160 sq rods, or 4,047 m^2. One acre is 0.4047 hectare.

ACRE-FOOT Usually applied to water volume or capacity of one foot depth covering one acre, or equivalent of 43,560 ft^3. One acre-foot equals 1233.5 m^3.

ACRS Advisory Committee on Reactor Safeguards.

ACS American Chemical Society.

ACTINIDES Elements with atomic numbers ranging from 89 to 103 (Table A-1). The elements in this series are radioactive and include the nuclear fuels and nuclear fertile materials. Also known as radioactive rare earth elements.

TABLE A-1
Actinide Elements (Radioactive Rare Earth)

Element	Symbol	Z, Atomic Number	amu, Atomic Weight
Actinium	Ac	89	227
Thorium	Th	90	232.038
Protactinium	Pa	91	231.0359
Uranium	U	92	238.03
Neptunium	Np	93	237.0482
Plutonium	Pu	94	244
Americium	Am	95	243
Curium	Cm	96	247
Berkelium	Bk	97	247
Californium	Cf	98	251
Einsteinium	Es	99	254
Fermium	Fm	100	257
Mendelerium	Md	101	256
Nobelium	No	102	254
Lawrencium	Lr	103	256

ACTINOMETER An instrument used for meteorological measurements of solar radiation, such as a pyrheliometer. The intensity of photochemically active radiation is measured by determining the extent of a chemical decomposition reaction by the incident radiation.

ACTIVATED CARBON A solid chemical compound consisting of 90-95% carbon used for adsorbing vapors and gases, particularly those affecting odor and taste. The material has 500-1000 square meters/gram of surface area. Organic vapors displace water present on the carbon. The carbon can be stripped of adsorbents with steam. Vapors are adsorbed approximately inversely to the ratio of the molecular volumes, inversely with temperature, and directly with pressure. Activated carbon may absorb a weight equal to the dry weight. Activated carbon is also called activated charcoal, active carbon, adsorbent carbon, decolorizing carbon, and vegetable carbon.

ACTIVATED CHARCOAL See Charcoal.

ACTIVATION ENERGY A constant whose value describes the effect of temperature on the rate of a reaction. The rate is proportional to

$$e^{-\frac{E_A}{kT}},$$

where E_A is the activation energy, T is the absolute temperature of the reactants, and k is Boltzmann's constant. The factor above is sometimes written

$$e^{-\frac{Q_A}{RT}},$$

where Q_A is the activation energy per mole and R is the gas constant.

ACTIVATION PRODUCTS Substances which are made radioactive by absorption of neutrons in a reactor.

ACTIVE MATERIAL The part of a material that is responsible for its characteristic action. For example, the active material in natural uranium fuel is the fissionable isotope ^{235}U.

ACTIVE POWER Power available from the wind, flowing water, ocean currents, etc., which could be converted to other forms for use.

ACTIVE SOLAR ENERGY SYSTEM A solar heating or cooling system which depends on external energy to circulate a working fluid and to operate controls connected to the system.

ACTIVITY The ratio of change from one to another condition or with time, such as activity coefficient, biological activity, chemical activity (electromotive series), radioactivity, and water activity.

ACTIVITY COEFFICIENT The ratio of effective and actual contribution of a substance to a phenomenon under consideration. For a gas the effective pressure is represented by the fugacity (escaping tendency), f, and the actual pressure, p. The activity coefficient is $\gamma = f/p$. For a gas, one means of determining f, and thus γ, is on the basis of the deviation of the volume of a real gas from an ideal gas. (Raoult's law and Henry's law.) The principle is also applied to a component in a solution, in which case a measure of nonideal behavior is represented.

ADIABATIC A thermodynamic process which occurs without a transfer of heat to or from an external source. Also see *k*.

ADSORPTION The process in which gases, liquids, or dissolved substances are held on the surfaces of solids. Also see *Absorption*.

ADVECTION In contrast to vertical flow of fluid resulting from a difference in gravity, known as convection, advection is the flow of air in a horizontal direction usually linked to or a part of the convection cycle (Figure A-2).

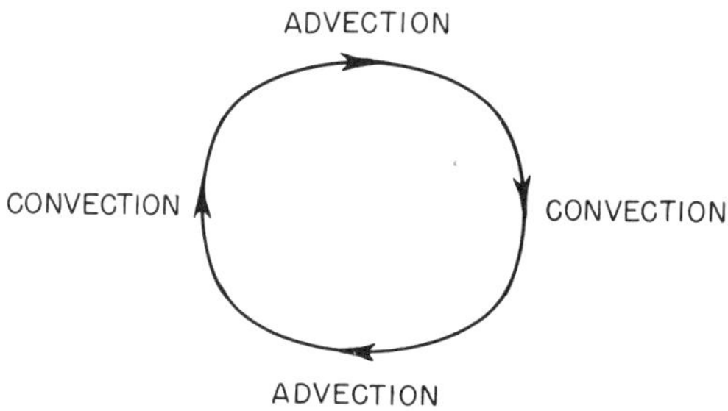

Fig. A-2 Advection.

AEC Atomic Energy Commission (replaced by NRC, in part).

AEE Association of Energy Engineers.

AEOLIAN ENERGY Energy produced from the wind.

AEROGENERATOR A general name applied to windmills, but in practice is more often used to identify a 2-3 bladed wind turbine or wind generator for producing electricity. Also see *Wind Energy; Wind Turbine; Windmill.*

AEROSOL Air- or gas-supported solid or liquid particles produced by natural and man-made activities, usually in the size range of approximately 0.1 to 1 micrometer. The larger materials tend to settle out and the smaller particles tend to agglomerate. The formation of aerosols which pollute the atmosphere, primarily from combustion processes, is a major concern. Also see *Particulates.*

AEROSPHERE The blanket of atmosphere surrounding the earth. Also see *Atmosphere.*

af Acre-feet.

AFB Atmospheric fluidized bed.

AFUP Alternative Fuels Utilization Program.

AGA American Gas Association.

AGE See Eon; Epoch.

AGGLOMERATION The process during which fine particles gather together into larger particles. The process is used to join small particles into agglomerates to reduce dust and to increase solubility.

AGR Advanced gas-cooled reactor.

AIChE American Institute of Chemical Engineers, 345 E. 47th St., N.Y., NY 10017.

AIF Atomic Industrial Forum.

AIME American Institute of Mining, Metallurgical and Petroleum Engineers, 345 E. 47th St., N.Y., NY 10017 (Figure A-3).

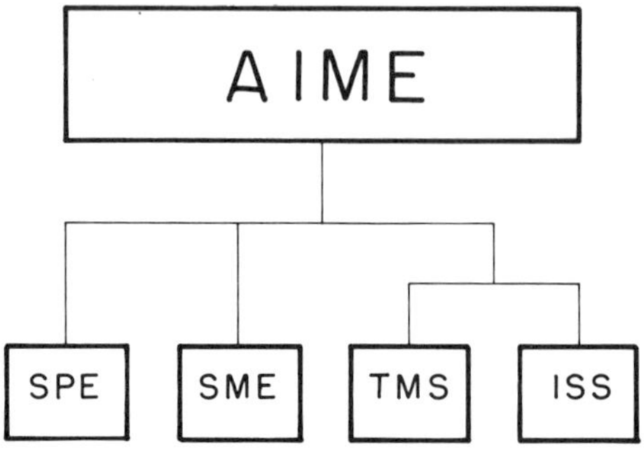

Fig. A-3 AIME.

AIR A mixture of gases and vapors that surrounds the earth. Dry air is a colorless and odorless mixture which at sea level consists of 78% nitrogen and 21% oxygen by volume (76.8% nitrogen and 23.2% oxygen by weight) plus several minor elements and usually some water vapor. Air is compressible and at constant temperature the density increases in proportion to the absolute pressure. The properties of air are summarized in Table A-2. Also see *Atmosphere*.

TABLE A-2
Air Properties

Properties	English Units	SI Units
Density at one atm. pressure	lb/ft^3	kg/m^3
0°C (32°F)	0.08065	1.2929
20°C (68°F)	0.07516	1.2050

AIR HORSEPOWER The power required to move air through a system of ducts or other resistances to air flow. The power which can be produced from air which is moving, such as from a compressed air source. Represented by ft-lb per min divided by 33,000. Also called fluid horsepower. Also see *Fluid Horsepower*.

AIR QUALITY INDEX (AQI) A means of identifying the general

quality of air which incorporates several factors, in which a number of 100 or less represents acceptable air from a health standpoint. Also see *Pollutant Standard Index (PSI)*.

ALBEDO The ratio of the reflected radiant energy to the incident (incoming) radiant energy for a surface generally applied to the earth's surface, where the approximate albedo values are: dark forest, 0.04; bright sand dune, 0.28; ocean, 0.05-0.06; fresh snow, 0.80-0.90; land surface, 0.12; wood, pine, 0.40; and earth, wet, 0.08.

ALCOGAS A fuel obtained by blending alcohol and gasoline. Other names for a mixture of alcohol and gasoline are: Agrol, Alcofuel, Alky-gas, Dynalkol, Gasohol, Methaline, Methline and Ethline, Discol, Koolmotor, Natolite, Petranol.

ALCOHOL Colorless chemicals of (OH) radicals attached to hydrocarbons, no more than one (OH) to a single carbon atom. Those with carbon C_1 to C_{11} are liquids and those with over 12 carbons are solids at room temperature. As the number of carbon atoms and the molecular weight increases, the boiling point decreases and the water solubility increases. The simplest alcohol is methanol (methyl alcohol), CH_3OH. The word alcohol is often used as a common name for ethanol (ethyl alcohol), C_2H_5OH. Absolute alcohol is pure (99-100%) alcohol. Denatured alcohol has added to it a substance, such as another alcohol for special purposes as for a solvent, making it (ethanol) unfit for human consumption, in contrast to neat alcohol which is not mixed with materials. Fuel alcohol is alcohol with approximately 80-95% purity (160-190 proof) in water and with other constituents from the process. Also see *Absolute Alcohol; Azeotropic; Denaturant; Ethanol; Gasohol; Hydrous Alcohol; Methaline; Methanol*.

ALGAE Non-flowering plants ranging in size from single cellular plants to large seaweeds which, unlike fungi, contain chlorophyl. Algae in the ocean are responsible for much of the photosynthesis. Algae utilize about 2% of available solar energy for production of carbohydrates by photosynthesis. Algae could become significant sources of energy as a cellulose product. Also see *Brown Algae; Kelp*.

ALGAE COAL Coal which has developed primarily from algae.

ALGAL SHALE Also known as bituminous shale. Also see *Bituminous Shale*.

ALICYCLIC When referring to hydrocarbons (CH combinations), a hydrocarbon in which the arrangement of carbons forms a closed ring or rings such as occur in crude oil. Also see *Aliphatic; Aromatic*.

ALIPHATIC When referring to hydrocarbons (CH combinations), a hydrocarbon in which the carbons are arranged in an open chain. Also see *Alicyclic; Aromatic; Hydrocarbon*.

ALKANES Also known commonly as paraffins, are saturated open chain aliphatic hydrocarbons, which have the general formula C_nH_{2n+2} including methane, CH_4; ethane, C_2H_6; propane, C_3H_8; butane, C_4H_{10} and pentane, C_5H_{12}. As the number of carbons increases, there is an increase in boiling point, density, heat or calorific content, and less quantity occurring naturally. Also see *Hydrocarbon Series*.

ALLIANCE TO SAVE ENERGY Alliance to Save Energy, 1925 K Street, N.W., Washington, DC 20006.

ALLOTROPIC Referring to any one of several forms of a material, as, for example, there are several allotropic forms of carbon such as charcoal, graphite, and diamond. Also see *Carbon*.

ALLOTROPY The property of a molecule or crystal that permits it to exist in more than one geometrical form. The properties of the different allotropic forms may be quite different. For example, the physical properties of graphite and diamond, two allotropic forms of carbon, are very different. Also see *Allotropic*.

ALPHA PARTICLE A helium nucleus given off by a radioactive material. Radiation materials are unstable because of an excess of protons. The stream of alpha particles emitted provides an ionizing radiation, known as alpha radiation, which has weak penetrating power.

AMBIENT Conditions, usually environmental, surrounding an activity, such as ambient temperature or ambient pressure or ambient humidity.

AMBIENT TEMPERATURE The temperature of the surrounding atmospheric air. Ambient air temperature changes could affect the operation of equipment, instruments, or controls. Compensating devices may be used to adjust for ambient air temperature changes to provide greater accuracy of measurement and control.

AMERICAN GAS ASSOCIATION (AGA) American Gas Association, 1515 Wilson Boulevard, Arlington, VA 22209. An industry association which provides educational and technical information specifically to its members and generally to the public.

AMERICAN GEOLOGICAL INSTITUTE (AGI) A technical organization concerned with subjects in geology, located at 5202 Leesburg Pike, Falls Church, VA 22041.

amu Atomic mass unit, atomic weight.

AMERICAN NATIONAL STANDARDS INSTITUTE (ANSI) 1430 Broadway, New York, NY 10018.

AMERICAN NUCLEAR SOCIETY (ANS) An international technical society of engineers and scientists interested in development of nuclear science and technology. The headquarters for ANS is at 555 N. Kensington Avenue, La Grange Park, IL 60525.

AMERICAN PETROLEUM INSTITUTE (API) A trade organization located at 1801 K Street, N.W., Washington, DC 20006.

AMERICAN SOCIETY FOR TESTING AND MATERIALS (ASTM) An organization located at 1916 Race Street, Philadelphia, PA 19103, devoted to developing testing procedures and standards for materials.

AMERICAN SOCIETY OF AGRICULTURAL ENGINEERS (ASAE) An engineering organization with mailing address PO Box 410, St. Joseph, MI 49085.

AMERICAN SOCIETY OF MECHANICAL ENGINEERS (ASME) An engineering organization located at 345 East 47th Street, New York, NY 10017.

AMERICANS FOR ENERGY INDEPENDENCE Americans for Energy Independence, 1500 Wilson Boulevard, Arlington, VA 22209.

AMORPHOUS A material which lacks an orderly arrangement of atoms and has the same physical properties throughout. Amorphous means without structure, i.e., lacking a crystalline structure. Examples are glass and pitch.

ANABOLISM A process in which energy is stored in chemicals as simpler elements are combined into more complex molecules during metabolism (Figure A-4). Also see *Catabolism*.

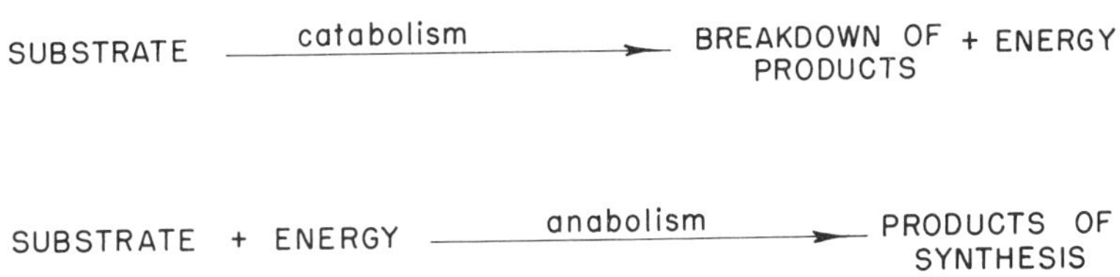

Fig. A-4 Anabolism.

ANAEROBIC DIGESTION; ANAEROBIC FERMENTATION The degradation or decomposition of organic matter by microorganisms in the absence of free or dissolved air (oxygen), usually in a water environment, under appropriate chemical and thermal conditions, producing methane, hydrogen sulfide, and carbon dioxide (Figure A-5). Methane is particularly useful as a source of energy.

ANERGY The quantity of energy not available for useful work, such as between the surrounding temperature and the absolute temperature,

Anergy = Energy - Exergy

Also see *Exergy*.

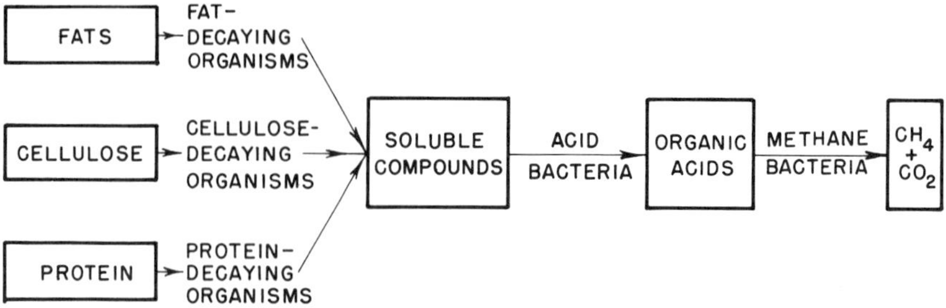

Fig. A-5 Anaerobic fermentation.

ANCILLARY ENERGY Energy external or in addition to that normally or usually a part of a process or system. It might also be called auxiliary energy.

ANGLE OF INCIDENCE See Incident Light.

ANGLE OF REPOSE The maximum angle from the horizontal which a bulk material will assume when piled, generally from 25 to 40°. The angle of repose, ϕ, for various materials is shown in Table A-3.

TABLE A-3
Angle of Repose, ϕ, Approximate

Product	Angle from horizontal, degree
Anthracite coal (hard)	27
Ashes	40
Bituminous coal (soft)	35
Cement	40
Coke	30-45
Corn, shelled, dry	27
Limestone, crushed and screened	37
Liquid	0
Malt	22
Rice	36
Sand, dry	25-35

Also see *Friction.*

ANGSTROM, Å A unit of length used particularly for wavelengths of electromagnetic radiation. One angstrom is 10^{-4} micron, 10^{-8} cm, or 10^{-10} m.

ANNIHILATION RADIATION The radiation that is produced when a particle and its antiparticle come together to annihilate each other. The most common example is the radiation produced by the annihilation of positrons by electrons. The basic reaction is

$$\underset{\text{electron}}{e^-} + \underset{\text{positron}}{e^+} \rightarrow \underset{\text{gamma rays}}{2\gamma}$$

where each gamma ray photon has an energy of 510 keV.

ANSI American National Standards Institute.

ANTHRACITE See Hard Coal; Malting Coal.

ANTHRACITIZATION The process of development of hard or anthracite coal from soft or bituminous coal, during which the carbon content increases and the volatile matter decreases. Also see *Anthracite Coal; Carbonization.*

ANTI-KNOCK, ANTIKNOCK The ability of a fuel to be used in an internal combustion engine without knocking following ignition. Lead additives to gasoline permit use of higher compression ratios. The octane number represents the ability to resist knocking, or antiknocking quality. Also see *Centane Number; Octane Number; Tetraethyl Lead.*

ANTIMATTER Matter composed of antiparticles. For example, the antihydrogen atom is composed of an antiproton and a positron. Antimatter does not exist in nature on earth because it would quickly be annihilated by the earth's matter. Also see *Antiparticle.*

ANTI-OXIDANTS, ANTIOXIDANTS An additive to gasoline to prevent oxidation of fuel in storage. Amine and phenolic compounds have been used for this purpose.

ANTIPARTICLE One member of a pair of fundamental particles with closely related properties. A particle and its antiparticle have the same mass but opposite electric charge. They can annihilate each other when they come into contact. The more common member of the pair is usually called the particle and the less common the antiparticle. Examples of pairs are the electron and positron, the proton and antiproton, the neutron and antineutron, and the neutrino and antineutrino. Also see *Antimatter.*

APA Alaska Power Administration. See *Power Administration.*

APCA Air Pollution Control Association.

API American Paper Institute.

API American Petroleum Institute.

API GRAVITY A scale for designating the density of oil,

$$\text{deg API} = \frac{141.5}{\text{SG }(60°\text{F}/60°\text{F})} - 131.5$$

APPARENT DENSITY The weight per unit volume of a material, including voids, which is a measure of the bulk of the material. Apparent density refers to the weight per unit volume whereas the apparent specific gravity denotes the specific gravity of a bulk volume, not including voids. Also see *Absolute Density; Bulk Density; Apparent Specific Gravity.*

APPARENT DIGESTIBLE ENERGY The gross energy in food consumed less the fecal energy. Also see *Fecal Energy; Metabolizable Energy.*

APPARENT SPECIFIC GRAVITY Denotes the specific gravity of a porous solid where the volume, V, excludes the permeable voids. The bulk specific gravity refers to measurements in which the volume, V, of the solid includes permeable and impermeable voids. Thus, bulk specific gravity is less in numerical value than apparent specific gravity. Also see *Bulk Density; Specific Gravity.*

AQI Air Quality Index.

AQUACULTURE (Also spelled aquiculture) The cultivation and production of plants and animals and other water-borne organisms in bodies of water, either inland or offshore. A specialized subject dealing with raising of plants and animals for human use or consumption in the large lakes and oceans is known as mariculture. Mariculture may be applied more broadly as the cultivation of marine organisms.

ARC - COAL An oil substitute made from finely ground coal developed by Atlantic Research Corporation. It is a heavy oil which will replace oils used in boilers and furnaces such as used for ships and utilities.

ARGON An inert gas making up about 1% of the atmosphere. Argon does not enter into combustion reactions. It is used as an insulator (or blanket) in nuclear reactors based on its inert qualities and heat transfer properties.

AROMATIC When referring to hydrocarbons, a hydrocarbon in which the arrangement of carbons form a closed ring or rings. The simplest aromatic hydrocarbon is benzine (C_6H_6) which results from refining petroleum and making coke from coal. The aromatic hydrocarbon is similar to the alicyclic hydrocarbon except that there are alternating bonds joining the carbon atoms. Also see *Alicyclic; Aliphatic.*

ASA American Standards Association.

ASAE American Society of Agricultural Engineers.

ASBESTOS A group of fibrous minerals composed primarily of magnesium silicate, which is heat-resistant. It is used for high temperature thermal insulation and electrical insulation. Particles of asbestos are considered carcinogenic.

ASCE American Society of Civil Engineers.

ASH The solid, generally inert material which remains after combustion, black to gray in color. Ash is also a product of volcanic eruption. Ash may be expressed on a moisture-free basis or as it exists after formation. Ash is produced from burning various products such as coal, wood, oil, and plants. Also see *Fly Ash.*

ASLB Atomic Safety Licensing Board.

ASLE American Society of Lubricating Engineers.

ASM American Society of Metals.

ASME American Society of Mechanical Engineers.

ASPHALT A naturally occurring pitch called mineral tar or mineral pitch. Also used to describe residues from refining of petroleum, as well are terms pitch, bitumen, and tar. Asphalt is widely used as a road pavement. Also see *Bitumen; Pitch; Sulfur-extended Asphalt; Tar.*

ASSOCIATED GAS Natural gas which is found with petroleum deposits, either dissolved in the oil, or in the cap above the oil. Also see *Natural Gas.*

ASSOCIATION OF ENERGY ENGINEERS Association of Energy Engineers, 4025 Pleasantville Road, Atlanta, GA 30340.

ASTHENOSPHERE See Earth; Mantle.

ASTM American Society for Testing Materials.

ATGAS A method of gasification of coal in which the coal is dissolved in molten iron and gas produced. Also see *Bi-Gas; Coed; Cogas; Gasification; Hygas; Lurgi.*

ATMOSPHERE The gaseous blanket surrounding the earth and other bodies of the universe. The atmosphere of the earth varies only slightly over the face of the earth but varies primarily in density with altitude. Sea level is taken as a datum plant for physical characteristics of the atmosphere. At sea level the composition of dry air is:

	By Volume, %	By Weight, %
Nitrogen	78.08	75.5
Oxygen	20.95	23.2
Argon	0.93	1.33
Carbon dioxide	0.03	0.045

The atmosphere also includes particulates, and other elements, such as neon, helium, and krypton, and compounds such as water vapor, hydrocarbons, and sulfur compounds.

The term atmosphere is used to designate the pressure of the atmosphere in which 1 atm is 1.032 bar, 14.696 psi, 101.325 kPa, 760 mm Hg, and 29.92 in Hg.

The atmosphere is divided into the lower atmosphere, 0-16 km, which includes the troposphere; and the middle atmosphere, 16-80 km, which includes the stratosphere and mesosphere; and the upper atmosphere, 80-400 km, which includes the ionosphere. Beyond the ionosphere are the thermosphere and the exosphere. Also see *Exosphere; Ionosphere; Ozone.*

ATMOSPHERIC GRADIENT See Gradient.

ATOM The fundamental unit of matter consisting of a nucleus and its surrounding electrons. The nucleus has a positive charge, Ze, where Z is an integer that is characteristic of the element and e is the electronic charge. There are Z negatively charged electrons that balance out the positive nuclear charge. Most of the mass of an atom is in the nucleus, but the atomic size is determined by the surrounding electrons.

ATOMIC ENERGY Also called nuclear energy. Energy that is released by a nuclear reaction such as fission or fusion. The energy is often called atomic energy although the energy involved is nuclear, not extranuclear. Also see *Nuclear Energy.*

ATOMIC ENERGY COMMISSION (AEC) A federal organization established by legislation in 1946, to oversee and direct the development of nuclear power. In 1974 the AEC was terminated and many of its responsibilities shifted to the Nuclear Regulatory Commission (NRC) and the Energy Research and Development Administration (ERDA), now Department of Energy (DOE). Also see *Nuclear Regulatory Commission.*

ATOMIC FISSION See Nuclear Fission.

ATOMIC INDUSTRIAL FORUM (AIF) An association of organizations interested in the peaceful and productive applications of nuclear energy. The Forum develops recommendations, sponsors study projects, and publishes material of interest not only for its members but for the general public. It is headquartered at 7101 Wisconsin Avenue, Washington, DC 20014.

ATOMIC MASS UNIT (amu) See Atomic Weight.

ATOMIC NUMBER Represented by Z, the number of protons in the nucleus of an atom, with all elements having a distinctive atomic number. The atomic numbers of those elements commonly associated with energy are as follows: Carbon, 6; helium, 2; hydrogen, 1; nitrogen, 7; oxygen, 8; radium, 88; sodium, 11; strontium, 38; sulfur, 16; and uranium, 92. Also see *Actinide; Atomic Weight; Protons*.

ATOMIC PILE A nuclear reactor, the term "pile" originated with the original nuclear reactors which were made of blocks of graphite and fuel "piled" into configurations that gradually approached criticality as they became larger.

ATOMIC REACTOR A nuclear reactor.

ATOMIC WEIGHT The relative weights of atoms of various elements usually expressed as atomic mass units (amu). The amu values usually are very close to whole numbers. The amu is defined as 1/12th of the mass of an atom of carbon 12, which has six protons and six neutrons in its nucleus. The atomic weights of those elements commonly associated with energy are as follows: Carbon, 12.01115; helium, 4.0026; hydrogen, 1.008; nitrogen, 14.0067; oxygen, 15.994; radium, 226; sodium, 22.9898; strontium, 87.62; sulfur, 32.064; and uranium, 238.03. Also see *Atomic Number*.

ATOMIZATION The process of breaking or dividing a liquid into very small drops or droplets. Atomizers are used for fuels and for drying or dehydration of liquid products in spray dryers. The total surface area of a unit volume available for heat and mass transfer is greatly increased as the size of drop is decreased.

ATOMIZER A device for producing droplets from a liquid, such as for atomization of fuels for a burner and for atomization of liquid products for spray drying. The atomized fuel is burned in a combustion chamber. The atomized product for drying is introduced in the drying chamber where heated air evaporates moisture from the atomized product and carries the moisture out of the dryer. Atomization of products is accomplished by a high pressure nozzle, a two-fluid (air or steam) atomizing nozzle, or a centrifugal disc (disk).

ATOMIZING BURNER Devices for producing droplets from fuel oil for combustion. Various types of burners for fuel oil are high pressure air atomizing, low pressure air atomizing, spray nozzle, rotary mechanical atomizing, and steam atomizing.

ATP Adenosine triphosphate.

ATTIC OIL Another name, not of preferred usage, for crude oil.

AVAILABLE ENERGY A term applied to the quantity of energy available for the process in question. The available energy is represented in a thermodynamic system by $(h - T_0 s)$ where h is the enthalpy, $T_0 s$ is the unavailable energy, and T_0 is the lowest temperature at which heat can be discarded.

AVAILABILITY FACTOR A measure of the maximum theoretical effiency of a thermal system. A high availability factor represents a high quality heat, or high availability of heat. It is represented by

$$\frac{T - T_0}{T}$$

where T is the absolute temperature of the heat supplied and T_0 is the temperature of the heat exhausted. Also see *Low Grade Heat.*

AZEOTROPE A given mixture of liquids which has a constant boiling point, for which the composition of the liquid does not change as it is converted to vapor.

B

B, b   Barrel.

BACKGROUND RADIATION   Radiation present from natural rather than man-made sources. Background nuclear radiations come from cosmic rays arriving from outer space and from naturally occurring radioactive materials present in the earth and in building materials. Background sound radiations are produced by the wind, animals, waterfalls, etc. Background light radiations at night are moonlight, starlight, and natural phosphorescence.

BAGASSE   The residue which remains after mechanical pressing of sugar cane. Bagasse has a moisture content in the range of 40-50% wet basis used for boiler fuel, and has a heating value of 18,500-21,000 kJ/kg (8000-9000 Btu/lb) of dry matter.

BAGHOUSE   A chamber for supporting and housing fabric filters which remove fine particles from air leaving a process. The filtered air is discharged to the atmosphere and the collected product is delivered to a hopper. The bags have a rating based on an airflow of 0.001 to 0.005 m^3/m^2s (1/5 to 1 cfm/ft^2) of fabric area. The bags may be vibrated periodically to remove the material. Fabric bags are made of natural fibers, such as cotton, and glass and plastic fibers. Also see *Filter*.

BAR   A unit of pressure equal to 1 million $dynes/cm^2$ which is 0.9868 atmosphere or 10^5 Pascal. A microbar is 1 $dyne/cm^2$.

BARGE   A large flat-bottom boat, usually pulled by a tug, used to transport coal, and grain, among other commodities. A river barge holds about 750 tons. A tank barge, as used for oil, may carry in the order of 80,000 bbl. Also see *Collier*.

BARK   The outer layer of a tree which can be used for fuel. The refuse of wet wood from a paper, lumber, or plywood operation. A variety of wet bark along with sawdust, butt ends, and culled pieces may be included. These materials can be used for fuel but a certain amount of pre-processing is desired to provide a material with consistent burning. Bark may contain as much as 80% moisture, wet basis. For wet bark there may not be enough heat available to support combustion. Dry bark has a heating value of about 21,000 kJ/kg (9000 Btu/lb). Mechanical presses can be used to reduce the moisture content from 80 to 65% wet basis. Stack or flue gases can be used for drying the pressed wood in bins. Wood waste for fuel is also called hogged fuel.

BARN A unit often used to express the size of a nuclear reaction cross section. One barn is a cross section equal to 10^{-24} cm^2.

BARREL Forty-two U.S. gallons (306 lb) or 35 imperial gallons. The barrel is the usual unit used to measure bulk quantities of crude oil and petroleum products for international trade. The amount of energy per barrel is approximately: Crude oil, 5.6×10^6 Btu; gasoline, 5.25×10^6 Btu; kerosine, 5.7×10^6 Btu; distillate fuel oil, 5.8×10^6 Btu; and residual fuel oil, 6.3×10^6 Btu.

BARRIER SHIELD A structure built around a source of nuclear radiation (e.g., core of a nuclear reactor) to absorb the radiation from it.

BARYON Any member of a family of particles which includes the neutron and proton, their antiparticles and a number of heavier members.

BASEBALL CONFIGURATION A magnetic field configuration used to confine a plasma. The name is derived from the fact that the electrical conductor used to produce the magnetic field is shaped like the stitching that joins the two halves of the cover on a baseball (Figure B-1).

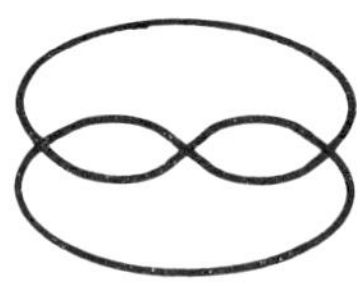

Fig. B-1 Baseball configuration.

BASELOAD POWER PLANT A facility for generating electricity which is designed to meet the maximum continuous demand. Additional demands are met by purchasing power as part of the network or grid and by peaking load generators.

BAT Best available technology.

BATCH A quantity of fuel, chemical, crop, or other substance placed into a bin or container on a repetitive basis specifically for treatment or processing. Used to describe a batch process in contrast to a continuous process.

BATEA Best available technology economically achievable.

BATTERY A device that produces an electrical voltage between two points called the terminals. When the terminals are connected to an electrical circuit, the battery delivers energy to elements of the circuit. A battery that can be recharged is called a storage battery. Also see *Energy Density*.

BAUMÉ Name of a hydrometer and scale for determining and designating the density of a liquid, for which 0° is the reading in water and 10° is the reading in a solution of 10% sodium chloride at 12.5°C (54.5°F).

BBEC Busbar energy cost.

BBL, Bbl, bbl Barrel(s).

b/cd, bpcd Barrels per calendar day.

Bcf Billion cubic feet.

BCR Bituminous Coal Research Institute.

BCT Best control technology.

BCURA British Coal Utility Research Association.

bd, B/D Barrels per day.

BDG Brewer's dried grain.

BDT Bone dry tons.

BEIR Biological effects of ionizing radiation.

BENEFICIATION The process of improving the grade of a mineral such as coal without involving chemical changes, for direct use or further processing. Beneficiation may involve sizing, sorting, and separation.

BENINUM Methane, CH_4, removed from petroleum which has been purified.

BENZENE A colorless liquid hydrocarbon, C_6H_6, having a ring structure obtained from coal and used as a chemical feedstock not to be confused with benzine. Benzene is a solvent for fat and is used in dry cleaning. The formation of six carbon atoms in a ring is called a benzene ring (Figure C-4). Also see *Benzine; Benzol.*

BENZINE A low-boiling point distillate of petroleum, a general term covering a mixture of hydrocarbons used for a fuel or solvent. Also see *Flash Point; Hydrocarbon Series.*

BENZOL Benzene with the impurities as when first produced. Benzol is further refined to produce benzene.

BEPS Building energy performance standards.

BERGIUS PROCESS A process for making gasoline from coal or cellulose utilizing high temperature and high pressure to produce a liquid which is treated with a catalyst (Figure B-2).

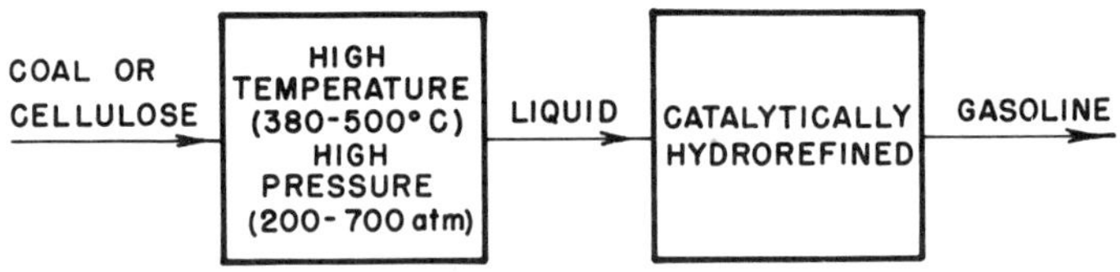

Fig. B-2 Bergius process.

BERYLLIUM A metallic element used as a neutron moderator by the nuclear industry. In combination with a source of high energy gamma rays, such as radium, beryllium becomes a source of neutrons.

BETA DECAY A process during which an electron (beta particle) is emitted from a nucleus at the same time that the nucleus spontaneously changes. An example is the beta decay of ^{60}Co during which the decaying cobalt nucleus changes to a nucleus of nickel. At the same time an electron, also called a beta particle or beta ray, is emitted. Often one or more gamma rays are emitted at the time of the beta deday.

Energy is released during beta decay. This energy is responsible for the afterheat that follows the shutdown of a nuclear reactor.

BETA EMITTER A material that undergoes beta decay and thus emits beta particles.

BETA PARTICLE, -RADIATION, -RAY Negative or positive electrons created during beta decay and emitted by the beta-emitting material.

BETZ LAW The maximum energy that can be extracted from the wind by a mechanical device is 59.3% of the energy of the wind.

BeV Billion electron volts. See *Electron Volt*.

BI-GAS (or BIGAS) PROCESS A process for producing methane from coal in a two stage pressure gasifier (Figure B-3). Coal and steam, with some oxygen added to the char, enter the first stage in which hydrogen is produced. The first stage is at or above 1370°C (2500°F). The volatile portion of the coal enters the second stage and is converted to methane at about 927°C (1700°F) at 100 atmospheres pressure. Gases leaving are treated by methanation to produce high-Btu gas.

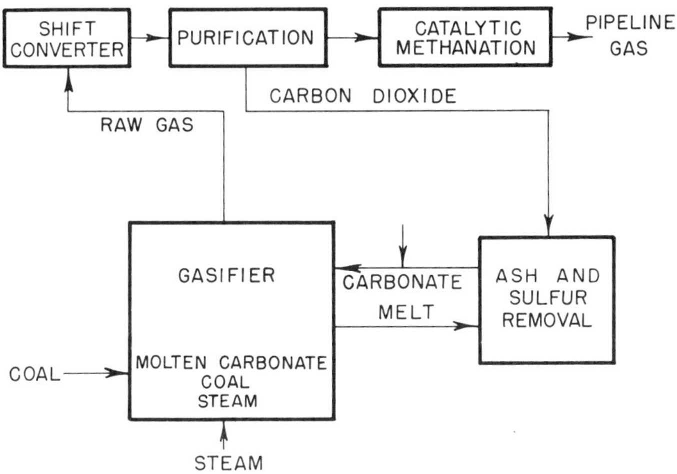

Fig. B-3 Bigas *(Hottel and Howard, 1971)*.

BIGIT A binary digit, also known as a bit, used as a measure of information.

BILLION A large number which has different values in various countries. In the United States of America and France, one billion is 1×10^9; in England and other Western European countries, one billion is 1×10^{12}. These numbers are also expressed as a thousand million and a million million, respectively.

BILLION ELECTRON VOLT (BeV) An amount of energy equal to 0.0016 erg, usually used to describe the energy of a single particle such as an electron or proton. Particles acquire high energies in machines called particle accelerators. Some large particle accelerators are: The Stanford Linear Accelerator (SLAC), the Fermilab Accelerator (FERMI), near Chicago; and the Bevatron at Brookhaven, New York. Particles with energies in the BeV range also arrive from outer space, in which case they are called cosmic rays. Also see *Particle Accelerators*.

BINARY CYCLE A combination cycle for operation of an engine that utilizes two separate cycles to convert energy from a primary heat source into mechanical energy. In one form, usually called a binary fluid cycle, or indirect cycle, the primary circuit carries energy from the heat source to a heat exchanger where it is transferred to the secondary that supplies a turbine or other thermo mechanical converter. In another form, often called a combined cycle system, the primary cycle also contains a turbine which converts part of the heat energy to mechanical form. The waste heat from this turbine then supplies energy to the secondary circuit through the heat exchanger.

BINDER Material used to hold small particles together, such as a binder for briquettes. Solid hydrocarbons, such as bituminous coal and lignite, are compressed into cubes, flat-cylinders, ellipsoids, and held together with a binder such as starch, tar, or asphalt. About 135 pounds of asphalt are used to produce a ton of briquettes. Sawdust may be pressed into briquettes. Also see *Briquettes*.

BINDING ENERGY (nuclear) The energy required to separate a nucleus into its constituent protons and neutrons (Figure B-4). The term is also applied to the energy required to remove a proton, neutron, alpha particle, or some other group of nucleons from a nucleus. In the latter case reference is made to the binding energy of the last proton, neutron, etc.

BIOCONVERSION The general designation which describes the conversion of energy by the use of biological mechanisms, microorganisms, or plants. The digestion of animal or human wastes by microorganisms to produce methane is an example. The production of carbohydrates by plants with solar energy is another example.

BIODEGRADABLE Capable of being decomposed by the action of living organisms, especially bacteria. Materials that are not biodegradable are especially undesirable as wastes because they remain in the environment for long periods of time.

BIOFOULING The growth of marine and water organisms on heat exchangers which decrease the rate of heat transfer.

BIOFUEL Fuel made directly from biological materials.

BIOGAS A product resulting from the digestion of organic materials, usually anaerobic, generally with a methane content of 55-70% and a carbon dioxide content of 30-45% (Figure B-5). Also called marsh gas and Gobar gas. Also see *Biopause*.

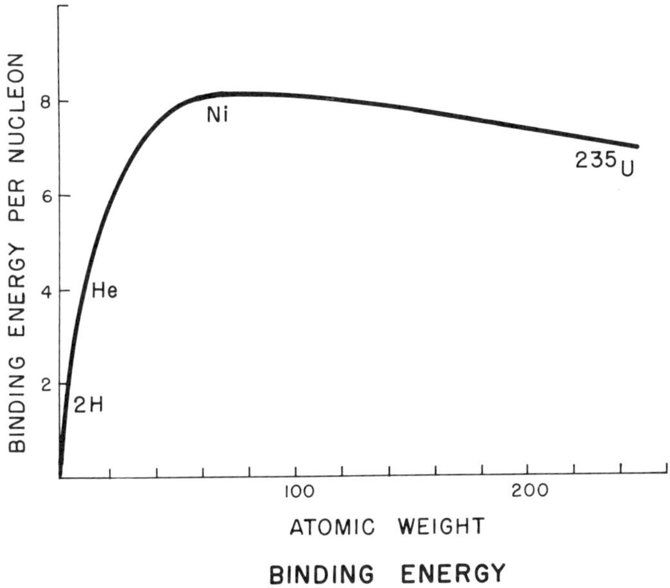

Fig. B-4 Binding energy *(Frauenfelder and Henley, 1974)*.

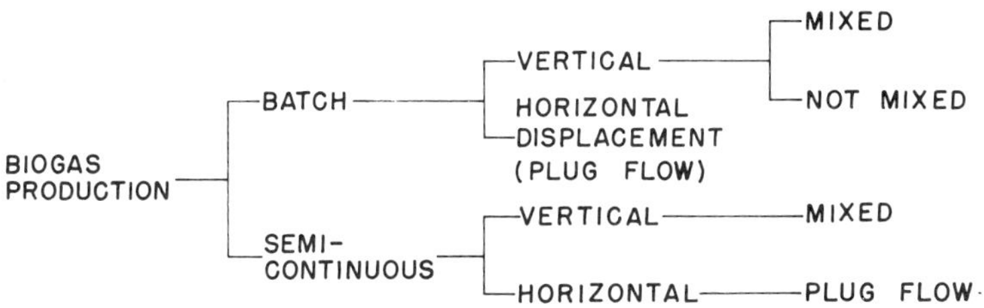

Fig. B-5 Biogas production approaches.

BIOGASIFICATION The process of gasifying a biological product by thermal treatment.

BIOGENIC A substance produced from biological action, such as biogenic methane, in contrast to abiogenic. Also see *Abiogenic*.

BIOLOGICAL OXYGEN DEMAND (BOD) A measure of both the biological and chemical oxygen demand of a pollutant or effluent, such as sewage. The substance is heated to kill organisms, then heated in the presence of an oxidizing agent, such as potassium dichromate, for an extended period, such as 5 days, at 20°C. Also see *Chemical Oxygen Demand.*

BIOLOGICAL SHIELD A structure built around the core of a nuclear reactor to absorb the radiations produced in it. The biological shield usually is designed to absorb enough of the radiation so that operating personnel can safely perform necessary tasks near the reactor but outside the shield.

BIOMASS All forms of plant and animal materials grown on land or in the water, and substances derived from biological growth, including animal, plant, and human wastes and residues consisting primarily of carbon, hydrogen, and oxygen (carbohydrates) and, to a minor extent, carbon and hydrogen (hydrocarbons). Also see *Cellulose; Energy Potential for Crop Systems; Lignite.*

BIOPAUSE The time span between the active growth phase and the senescence or slow growth phase in the biogas production curve (Figure B-6).

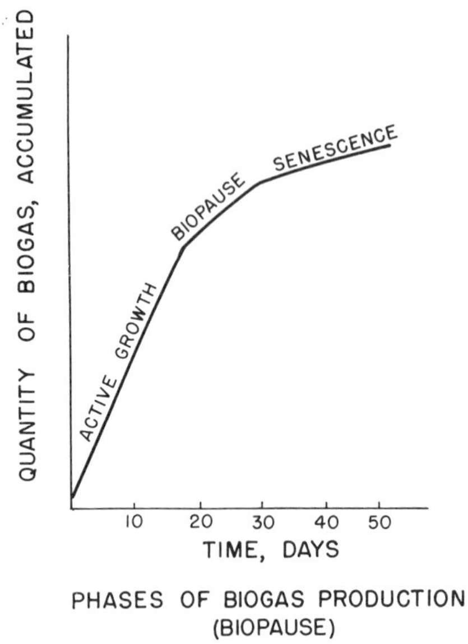

Fig. B-6 Biopause *(Maramba, 1978).*

BIOPHOTOLYSIS The decomposition of water into oxygen and hydrogen using the energy of light and biological catalysts.

BIOSPHERE The oceans, upper land mass, and lower atmosphere of the planet in which life exists. Also see *Geosphere*.

BIO-SYNFUEL Synthetic fuel produced from biological materials.

BIO-SYNTHOIL Synthetic oil produced artificially from biological materials.

BIOTOPE An area on the land or in the sea populated by a community of organisms with similar characteristics.

BIT See Bigit; Byte.

BITUMEN A general term applied to a mixture of hydrocarbons including crude petroleum, asphalt, pitch, and tar, which can occur naturally or can be obtained by distillation from crude oil or coal. Also see *Asphalt; Elkerite; Pitch; Tar*.

BITUMINOUS A general term referring to a solid or heavy liquid hydrocarbon with a high carbon content which will burn. Also see *Bituminous Coal; Bituminous Sand; Bituminous Shale*.

BITUMINOUS COAL A solid fossil hydrocarbon mineral, high in volatile matter also called soft coal. May be classified as high volatile and low volatile coal. Also see *Coal; Coal Classification; Hard Coal; Soft Coal*.

BITUMINOUS SAND Another name for oil sand or tar sand. It is a sand which contains a heavy asphaltic crude oil, requiring special means to separate. Also see *Oil Sand; Tar Sand*.

BITUMINOUS SHALE Another name for oil shale, black shale, coal shale, kerogen shale, algal shale, organic shale, lignitic shale, carbonaceous shale, tasmanite, etc. It is a sedimentary material containing a combustible solid, organic matter, called kerogen. Also see *Kerogen; Oil Shale*.

BLACK LUNG An occupational disease of coal miners caused by inhalation of dust, known as coal worker pneumoconiosis.

BLACK SHALE See Bituminous Shale.

BLACK WATER The water coming from coal separation plants which contains coal and rock particles.

BLANKET (nuclear) A layer of fertile material around the core of a nuclear reactor. The fertile material is partially converted into nuclear fuel by neutron absorption as the reactor operates.

BLAST-FURNACE GAS A low-Btu gas, 90 Btu/ft^3, a by-product of iron manufacture, which results from incomplete combustion of coke.

The characteristics of blast-furnace gas are: Specific gravity, 1.02 (compared to air); nitrogen, 60%; carbon monoxide, 27.5%; carbon dioxide, 11.5%. Blast-furnace gas is used primarily for heating applications. Also see *Carbon Dioxide; Carbon Monoxide; Low-Btu Gas; Nitrogen.*

BLEVE Boiling liquid-expanding vapor explosion.

BLM Bureau of Land Management.

BLOWDOWN Portion of steam or water removed at regular intervals from a boiler, cooling tower, etc., to prevent accumulation of dissolved and suspended materials.

BLOWER The terms fan and blowers are often used interchangeably. However, radial flow fans are commonly referred to as blowers and axial flow fans are commonly called fans.

BLWR Boiling water reactor (same as BWR).

BNFL British Nuclear Fuels, Limited.

BNL Brookhaven National Laboratory (NY)

BOARD FEET See Thousand Feet Board Measure.

BOD Biological oxygen demand.

BOE Barrels of oil equivalent (5.8 million Btu's).

BOILER A device for producing steam or hot water, consisting of an enclosed container in which water is heated under pressure. Either the water is circulated in tubes and heated, known as a water-tube boiler, or the hot gases are circulated in tubes to heat the surrounding water, known as a fire-tube boiler. Accessories and controls are used to provide the proper pressure and temperature, flow of gases and water, and fuel.

BOILER EFFICIENCY The ratio of energy increase of the water in the boiler to the heat supplied by the fuel, multiplied by 100 to obtain the efficiency in percent.

$$\text{Efficiency} = \frac{G_a(h_1 - h_2)}{G_f(H)}$$

where

G_f = weight of fuel burned, kg/hr or lb/hr
H = calorific heating value of fuel, kJ/kg or Btu/lb
G_a = evaporation rate kg/hr or lb/hr
h_2, h_1 = enthalpy of steam entering and leaving the boiler, respectively, kJ/kg or Btu/lb

BOILER HORSEPOWER (BHP)   Rating of a boiler based on the equivalent of evaporation of water at the rate of 34.5 lb/hr for one BHP at atmospheric pressure. In SI units one BHP = 9.8 kW.

BOILING   See Evaporation.

BOILING LIQUID-EXPANDING VAPOR EXPLOSION (BLEVE)   A violent rupturing of a pressure vessel which can occur when a partially filled pressure vessel is exposed to a fire.

BOILING WATER REACTOR (BWR)   In a boiling water reactor the cooling water flowing through the reactor core boils and converts to steam before it leaves the core (Figure B-7). The steam then passes through a turbine which drives an electrical generator.

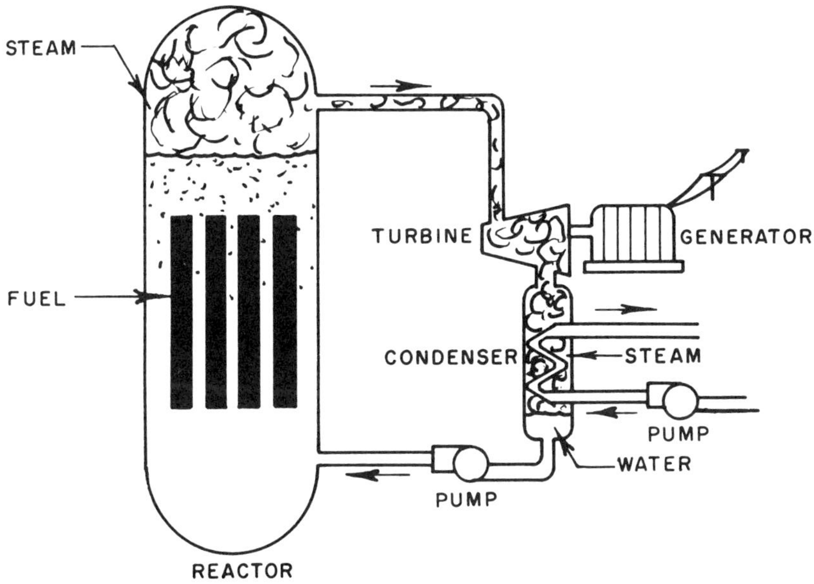

Fig. B-7   Boiling water reactor.

Many commercial nuclear power plants use the boiling water reactor concept.

BOM   Bureau of Mines.

BONNEVILLE POWER ADMINISTRATION (BPA)   A government agency which markets electric power generated in the Pacific Northwest primarily by projects built by the Corps of Engineers and Bureau of Reclamation.

BPA is now a part of the DOE, having formerly been in the USDI. Also see *Power Administration.*

BOND ENERGY The energy required to break a chemical bond. A typical bond energy is one electron volt. Bond energies are sometimes stated per mole of bonds broken. A typical value is 40 kilojoules per mole of bonds.

BOOSTER Most commonly a pump in a fluid stream to reduce the head on an existing pump or to increase the flow into an existing pump. In electric power systems a booster generator may be a synchronous generator to boost or buck converter d.c. voltage.

BOPD Barrels of oil per day.

BPA Bonneville Power Administration. Also see *Power Administrations.*

BPDE Barrels per day equivalent.

BORON A non-metallic element, one of whose isotopes, ^{10}B, is a strong absorber of thermal neutrons. Boron has often been used as part of the material in the control rods of nuclear reactors.

BOTTLED GAS Liquefied petroleum gases, butane and propane, made from petroleum, and stored under pressure in steel tanks. When made specifically for fuel, propane is the principal gas. Also see *Butane; Propane.*

BOTTOMING CYCLE A cycle or process added to a conventional energy cycle to extract additional power and improve the efficiency of a system. The use of the steam normally rejected from a conventional power plant to provide additional power (at a lower temperature) is an example of the bottoming cycle. Also see *Topping Cycle.*

BOTTOMS The residues of a distillation process that did not evaporate.

BRAKE HORSEPOWER The useful power available at the shaft of a power device, in which one horsepower is 33,000 ft-lb/min. The adjective, brake, is derived from the use of a Prony brake friction-loading device to determine the horsepower on a shaft. Also see *Horsepower; Power.*

BRANCHED CHAIN See Decay Chain; Hydrocarbon Series.

BRAYTON CYCLE Also called the Joule cycle, the Brayton cycle (Figure B-8) consists of two isentropic (1-2) and (3-4) and two isobasic (constant pressure) relationships (2-3) and (4-1), of particular applicability to gas turbine operation, with continuous operation and high gas throughput.

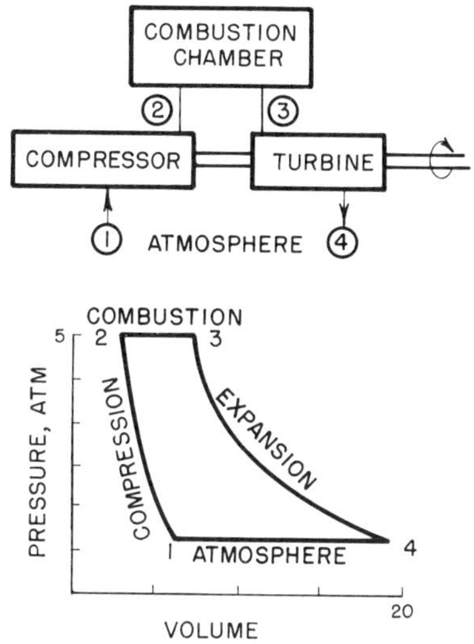

Fig. B-8 Brayton cycle.

BREEDER REACTOR A nuclear reactor that produces more fuel than it consumes. The most common type of breeder reactor produces the nuclear fuel ^{239}Pu from ^{238}U, the most common form of uranium. The transformation occurs in the core of the reactor where the nuclear energy also is released in the form of heat. Fuels that produce energy through nuclear fission include ^{239}Pu and ^{235}U. One form of breeder reactor is the liquid metal fast breeder reactor (LMFBR), which uses the liquid metal, sodium, as a coolant. The primary coolant (Na) exchanges heat to a secondary loop (Na) which in turn supplies heat to a steam generator. The remainder of the system consists of a conventional turbogenerator. Also see *Liquid Metal Fast Breeder.*

BREWER'S DRIED GRAINS (BDG) The dried extracted residue of barley malt or of a mixture with other cereal grain or grain products. These residues can be used for animal feed after alcohol has been extracted.

BRIGHT COAL In contrast to dull coal, a jet-black, compact constituent in coal. The coal is derived from woody plant tissues such as roots, branches, stems, and limbs.

BRIQUETTE; BRIQUET Compressed fines of coal or charcoal primarily for fuel. A binder of pitch or asphalt is commonly used. Starch may be used as a binder for charcoal briquettes. Also see *Binder.*

BRITISH THERMAL UNIT (Btu) A unit of energy equal to the heat energy required to raise the temperature of a pound of water through one degree Fahrenheit at 4°C (39.2°F). One Btu is equal to 0.252 kg-cal (Cal), 252 cal, 1055 Joules, 1.055 kJ, and 778.2 ft lb.

BROWN ALGAE A group of greenish yellow to brown plants which grow in the cooler waters of the world. The color is given more by carotenes than the chlorophyll. In addition to kelp, other plants such as rockweeds and gulfweeds are included. Also see *Algae*.

BROWN COAL Another designation for lignite, a carbonaceous material between sub-bituminous coal and peat. A further classification is hard brown coal, closest to sub-bituminous coal, and soft brown coal, closest to peat. Also see *Coal Classification; Lignites*.

BROWNOUT An alteration of normal electrical service during which the voltage supplied is reduced in order to save energy at a time of peak demand when the supply system cannot meet the full load.

b/sd Barrels per stream day.

Btu, BTU British thermal units.

BTX Benzene, toluene, and xylene.

BUBBLE CHAMBER A closed chamber containing a liquid, such as hydrogen, used in research on the properties of nuclear particles. In operation the pressure on the chamber is reduced suddenly leaving the liquid superheated. Under these conditions the liquid boils first at special sites including the tracks of ionizing particles which are moving through the chamber. The local boiling along a track leaves a trail of bubbles which can be observed and photographed through one or more sides of the chamber, which are made of transparent material. Bubble chamber photographs showing such tracks are studied to investigate properties of the particles and/or their interactions with the liquid inside the chamber.

BUG BATTERY A fuel cell or battery in which the flow of electrons is caused by growth of small "bugs" or microorganisms.

BULB-TURBINE A water turbine for hydroelectric power production designed especially to operate with the flow of water in either direction. An enlargement or "bulb" in the conduit is a physical feature of the unit. Water flow may be vertical or horizontal through the turbine. The bulb-turbine might be used for pumped-storage systems, tidal power plants, or other locations which might require switching input and output. Also see *Hydraulic Turbine; Pumped Storage*.

BULK DENSITY The gross or apparent density of a material which is the mass per unit of gross volume including void space. The bulk density of materials, ± 10%:

Cement	94-100 lb/ft^3	1505-1602 kg/m^3
Clay	63	1009
Coal, lump	50	801
Sand	90-105	1442-1682
Wood	20	320

BULKHEAD A partition of wood, concrete, or metal usually at the end of a container used for storage or processing to retain the products.

BULL KELP Included in the brown algae, a plant which is found on the Pacific coast of North America extending from California to Alaska. Also see *Brown Algae; Kelp.*

BUNKER A bin or vessel for storage, with a hopper bottom, usually for storage of solids.

BUNKER COAL A term applied to coal used for an ocean-, sea-, or lake-going vessel.

BUNKER OIL Any of a variety of heavy fuel oils made by stabilizing residual oil resulting from cracking of crude oil, especially used for sea-, ocean-, or lake-going vessels.

BURNUP FIGURE OF MERIT A figure of merit that indicates the amount of thermal energy released from nuclear fuel during the time that it remains in a reactor. The most common unit is megawatt-days per (adjacent) ton of fuel.

BUTANE A hydrocarbon gas, C_4H_{10}, in the paraffin series. Butane is recovered from natural gas or can be manufactured from petroleum. Butane can be liquefied under pressure at room temperature, 26.3 psia at 60°F, about one-fourth of the pressure for propane. Butane has a high heating value of 103,000 Btu/gal, or 21,200 Btu/lb. Also see *Methane; Propane.*

BUTANOL The alcohols with the molecular formula of C_4H_9OH which may have four different structures, the most common of which are 1-butanol (n-Butyl alcohol) and 2-methyl-1-propanol (isobutyl alcohol). Originally manufactured primarily from corn starch and molasses, and most recently made synthetically from ethanol or acetylene.

BWR Boiling water reactor.

BYTE A word used in digital computers representing a grouping of binary digits (bits), usually eight bits per byte. Older usage was for five bits per byte. Computers are also available which use 16 bits or 32 bits per byte.

1 kilobyte = 1024 bytes

A bit is also known as a bigit.

C

C$_4$-PLANTS Plants which are efficient photosynthetic converters, especially at high temperature and under intense light, as represented by sugar cane, sorghum, millet, and corn. These plants have a unique leaf anatomy which includes two cell layers concentrically arranged around the vascular tissue. The initial carbon dioxide fixation occurs in the outer layer of cells (mesophyll) and results in 4-carbon acids. The acids are transported to the inner layer of cells (bundle sheath) and are decarboxylated. CO_2 is then fixed in the Calvin cycle (a cycle common to all plants). C$_4$-plants are also described as having the Hatch and Slack pathway. Also see *CAM-plants*.

CADMIUM A metallic element used as a thermal neutron absorber in the nuclear industry.

CADMIUM RATIO Ratio of the activation of a foil irradiated in a neutron flux (nuclear reactor) to the activation of the same foil in the same flux when the foil is covered by cadmium. Cadmium is a strong absorber of thermal and slightly epithermal neutrons. The action of the cadmium cover is to filter out thermal and slightly epithermal neutrons so that the foil inside is not subjected to irradiation by these neutrons. Thus, the cadmium ratio is a measure of the activation produced by the neutron flux to the activation of the foil by neutrons with energies above the cadmium cut off. When a given foil and cadmium cover are used for a series of experiments in various neutron fluxes, the technique becomes a means of measuring how much of the flux is thermal.

CAGPL Canadian Arctic Gas Pipeline, Ltd.

CAISSON A temporary chamber built in water usually in the form of a cylinder or box, constructed of reinforced concrete, metal or wood, or combination, built to exclude water, to build footings, piers, foundations, etc. Caissons can be built so that they can be pressurized to help exclude water. Also see *Cofferdam*.

CALCINATION The process of decomposing carbonates by heating.

CALORIC A word no longer used. Previously used to represent the 'substance' of heat which was thought to move from one body to another. Some authors have mistakenly used the term caloric in place of calorific for the heat content of a substance. Also see *Calorific Value*.

CALORIE The quantity of heat required to raise the temperature of one gram of water from 3.5 to 4.5°C (38.3 to 40.1°F) known as the small calorie. The normal calorie is defined as the quantity of heat required to raise the temperature of one gram of water from 14.5 to 15.5°C (58.1 to 59.9°F). The large calorie written with a capital C to distinguish a Calorie, 1000 small calories, is also called Kilocalorie (kcal) or kilogram-calorie as contrasted to calorie or gram-calorie. Kilocalorie (kcal) is the preferred usage. A megacalorie (Mcal) is equivalent to 1000 kcal or 1 million calories. The mean calorie is defined as 1/100 of the amount of heat required to raise one gram of water from 0 to 100°C (32 to 212°F); 252 calories equal 1 Btu. One calorie equals 4.187 joules.

CALORIFIC VALUE The heat available in fuel by burning or in food by digestion, in calories (gram-calories) per gram, kJ/kg, or Btu/lb. The calorific value of fuels by combustion are commonly referred to as heating values. General calorific values for foods in calories/gram are: Cheese, 4-5; eggs, 1.5; milk, 0.7; meat, 1.1; bread, 2.5; cereals, 3.5-4.0; vegetables, 0.5-1.0; fruits, 0.2-1.0; nuts, 6.0-7.5. A 70 kg (154 lb) mature person has a basic minimum requirement of 1700 Cal/day and when involved in light work, 2300 Cal/day.

Calorific values (gross or net) of fuels are:

	Gross value, Btu/lb
Natural gas	26,000
Propane	21,500
Butane	21,200
Kerosine	20,000
Gas oil	19,500
Light fuel oil	18,750
Heavy fuel oil	18,500
Anthracite coal	15,000
Bituminous coal	12,000
Coke	12,000
Bagasse (dry)	8,000
Wood (dry)	7,000
Peat (50% water)	5,000
Wood (wet)	5,000

Conversion: *1 Btu = 1.0551 kJ = 0.2520 kcal = 252 cal.*

Calorific Values of Gases

	MJ/m^3	Btu/scf
Butane	92.5-101.5	3100-3400
Propane	71.7-77.6	2400-2600
Refinery gas	35.8-71.7	1200-2400
LNG	29.8-34.9	1000-1170
Natural gas	29.1-35.2	975-1180
SNG	29.0-29.4	970-985
Coke oven gas	14.6-16.7	490-560
Carbonated water gas	14.2-17.2	475-575
Producer gas	3.7-4.7	125-160
Blast furnace gas	2.1-2.8	70-95

Conversion: *1 Btu/ft^3 = 0.029856 MJ/m^3*

CALORIMETER A device for determining quantity of heat. When a given batch of substance is burned in the device to complete combustion, Btu/lb or kJ/kg, it is a bomb calorimeter; when it measures the heat in steam or heat of reaction it is known as a steam calorimeter; and when it determines continuously the heat content of flowing gas, Btu/ft^3 or kJ/M^3, it is known as a gas calorimeter. ASTM has established standards and procedures for determining the heat content, called calorific value. Also see *Carbon Cycle; High Heat Value.*

CALVIN CYCLE See C_4 plants; Carbon Cycle.

CAM Crassulacean acid metabolism (desert plants).

CAM-plants Crassulacean acid metabolism, a group of slow-growing plants in which most of the carbon dioxide is fixed in the dark rather than in light, such as in the pineapple and cacti plants. The main characteristic of the plants is a diurnal flux in acid content (malic) with high concentrations at night and less in the day. These plants utilize a system similar to C_4-plants to carboxylate and decarboxylate; however, the process takes place in one cell type rather than two. These plants are efficient consumers of water and thrive in arid and semi-arid areas. The efficient use of water occurs as a result of limited to no opening of stomates in the light. Also see *C_4-plants.*

CANDELA (cd) A SI basic unit of luminous intensity. The candela is the luminous intensity of a surface of 1/600,000 m^2 of a blackbody at the temperature of freezing platinum under a pressure of 101 kN/m^2. Luminance is a derived quantity, cd/m^2. Also see *Lumen.*

CANDLE COAL Also called Cannel Coal.

CANDLEPOWER A standard of illuminating power defined as that produced by a sperm candle which burns at 105.5 grains/hr (U.S.A. standard). For present usage, also see *Candela; Lumen.*

CANDU REACTOR A type of thermal reactor, developed in Canada, that uses heavy water to moderate (slow down) the fast neutrons produced by nuclear fission. There are no CANDU reactors operating in the U.S.A., but the system is commercially successful elsewhere.

CANNEL COAL An old coal high in volatile material derived from small pieces of plants and algae and spores. Synonyms for this coal include: Candle coal, kennel coal, and parrot coal. Cannel shale is a name sometimes given to oil shale. Also see *Bituminous Shale*.

CANYON (nuclear) A heavy shielded area in a fuel processing facility.

CAPACITY The quantity of throughput of an apparatus, machine, device, or process, which can be used for a rated load or maximum load and for either batch or continuous process.

CAPACITY FACTOR A number usually less than one or less than 100% which is the ratio of the energy generated by a power plant divided by the energy generated if the plant had operated at 100% of rated capacity over the same time.

CAPITAL ENERGY The energy stored in materials in the earth which is potentially available for use.

CAPTURE (nuclear) A nuclear reaction in which a particle is absorbed by a nucleus and no particle is emitted. De-exitation of the compound nucleus occurs by gamma ray emission.

CARBOHYDRATE Includes starches, sugars, and celluloses identified by the fact they all contain carbon, hydrogen, and oxygen, most of which are produced by green plants. The general formula for many but not all carbohydrates is $C_x(H_2O)_y$.

CARBON A chemical element essential to all plants and animals, and an important component in the combustion process for hydrocarbons and carbohydrates. The carbon fixed each year as a result of photosynthesis is approximately 10 times the amount involved in energy use. Also see *Activated Carbon, Carbon Cycle; Combustion*.

CARBON CYCLE The flow of carbon in living matter in which carbon dioxide is fixed by photosynthesis in green plants from organic nutrients and is ultimately returned to the inorganic state by respiration and decay (Figure C-1). The carbon cycle also refers to a cycle of thermonuclear reactions in which four hydrogen atoms synthesize into a helium atom with the release of nuclear energy and which is considered to be the source of most of the energy from the sun and stars. Also known as the Calvin cycle. Also see *C_4-Plants; Carbon Dioxide*.

CARBON DIOXIDE A colorless and odorless gas (CO_2) at atmospheric conditions which is a product of combustion of carbon-containing

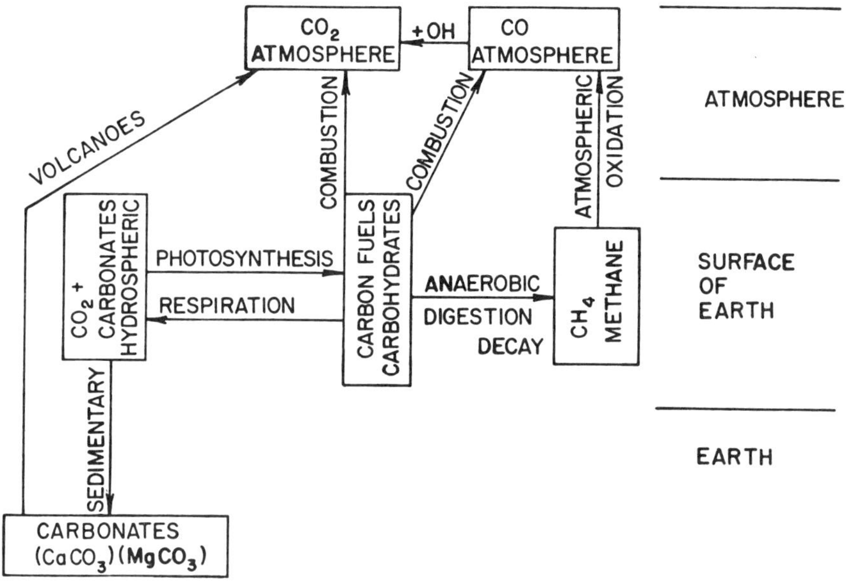

Fig. C-1 The carbon cycle.

fuels. The amount of carbon dioxide in the product of combustion is one measure of the efficiency of combustion. The carbon dioxide level of flue gas is about 21% for complete combustion of pure carbon in air. The flow of carbon dioxide is shown in Figure C-2. Also see *Carbon Cycle; Combustion*.

CARBON DIOXIDE ACCEPTOR See Consolidated Synthetic Gas (CSG).

CARBON MONOXIDE A highly poisonous colorless gas (CO) which is a product of combustion. Exhaust systems must be properly sealed and ventilated for safety of workers in the area. Incomplete combustion gives carbon monoxide and carbon dioxide plus water. Also see *Carbon Dioxide; Combustion*.

CARBONACEOUS Materials rich in carbon, particularly carbohydrates, which can be used to make fuels. Carbonaceous materials yield three times as much methane gas as proteinaceous materials, but in a longer time.

CARBON/HYDROGEN RATIO (C/H Ratio) A method of classifying carbon fuels based on the ratio between the carbon and hydrogen present, usually by weight. A coal which contains 75% carbon and 5% hydrogen has a C/H ratio of 15. Most bituminous coals have a C/H ratio of 14-17, and anthracite coals, 24-29. To convert coal to pipeline gas (CH_4), which has a C/H of 3 (75% carbon and 25% hydrogen), considerable hydrogen must be added which is the usual practice, or carbon

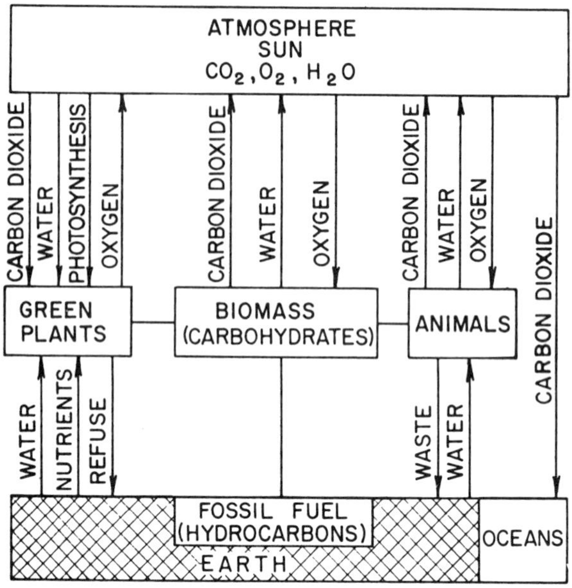

Fig. C-2 Carbon dioxide flow *(Hall, 1981)*.

must be eliminated. Also see *Hydrogen/Carbon Ratio*.

CARBONIFICATION The process of change in coal or formation of coal related to weight of earth over material, temperature, and pressure on the material. Generally there is an increase in carbon content during these changes. Also see *Coalification*.

CARBONIZATION The process of decomposition of a solid containing carbon such as coal or biomass into solid, liquid, and gaseous products by heating, generally in the range of 700 to 900°C. Also called pyrolysis, destructive distillation, or volatization.

CARBON/NITROGEN RATIO (C/N Ratio) A method of classifying materials on the basis of ratio of weight of carbon to nitrogen, for conversion of organic matter to biogas using bacteria in which nitrogen is needed for the life processes. It is generally accepted that the weight of nitrogen should be about 1/15th to 1/30th the weight of carbon, thus a C/N of 15-30 is desired, depending on the microorganism. Either too much or too little nitrogen can be detrimental to the growth of microorganisms. The C/N of straw is 51:1 and of paper is 400:1; thus, nitrogen needs to be added so that bacteria can digest these materials.

CARBONACEOUS SHALE See Bituminous Shale.

CARBONOUS A plant or animal material which resembles carbon in physical characteristics or appearance.

CARBURETTED WATER GAS A medium-Btu gas, approximately 500 Btu/ft^3, made by upgrading water gas with the addition of a hydrocarbon oil. Also see *Water Gas*.

CASINGHEAD GAS The gas in an oil reservoir which is dissolved in or associated with the oil and which may be removed from the oil well borehole, or the well casing. Also known as oil well gas.

CATA See Kata.

CATABOLISM A process in which energy is released by cellular breakdown during metabolism when complex chemicals are broken into simpler components, the opposite of anabolism. Also see *Anabolism*.

CATALYST A chemical which when present in a chemical reaction is not consumed as a part of the molecular structure of interaction, but is capable of changing the rate of that reaction. Metal catalysts are often used in gasification processes. Some of the metal ions in solid wastes serve as a catalyst in the formation of gases during a gasification process.

CATALYTIC CRACKING The process of breakdown of chemicals into high boiling point liquids accompanied by the production of large quantities of gaseous hydrocarbons in the presence of a catalyst usually at high temperatures.

CATALYTIC REFORMING A secondary refining process in which a catalyst is used to change the product. Catalytic reforming is used to make gasoline from the distillate of petroleum which is produced at 70°-190°C (158°-375°F). The octane number of the distillate is raised from 40 to 95, using a platinum catalyst, about 0.5%, at 490°-540°C (915°-1000°F), and at a pressure of 10-30 atmospheres. Also see *Catalytic Steam Reforming; Reforming*.

CATALYTIC STEAM REFORMING A process for producing hydrogen from a hydrocarbon, usually natural gas. In the process steam and natural gas are heated to 594°C (1100°F) or higher, a catalyst is introduced, hydrogen is produced and carbon dioxide, carbon monoxide, and hydrogen sulfide are removed. Also see *Catalytic Reforming; Reforming*.

CBO Congressional Budget Office (U.S.A.).

CDF Coal derived fuel.

CDL Coal derived liquid.

CEA Commissariat a l'Energie Atomique.

CELLULOSE A carbohydrate $(C_6H_{10}O_5)n$, a large polymer which has the same general formula as starch, with wood being the principal source. On complete hydrolysis cellulose yields sucrose. Cellulose may contain as many as 3000 glucose units to form a molecule. For cellulose, n may be 5000, as for cotton, and by comparison, n may be 1000 for starch. The woody parts and cell walls of plants and textile fibers, cotton and linen, are the major sources of cellulose. Cellulose, along with xylan, is a principal constituent in the cell walls of green plants. The molecular weight of cellulose is 16,000 to 2,000,000. The cellulose molecules are not digested by humans and non-ruminant animals.

°CELSIUS (°C) A unit representing temperature on a scale of 100 units between the freezing point (0°) and the boiling point (100°) of water at standard sea level atmospheric pressure which compares to 180 units on the Fahrenheit scale between 32°F and 212°F or 180 units. Formerly called Centigrade.

Celsius (°C) conversion to Kelvin (°K) and Fahrenheit (°F)

-273°C	0°K	-459.4°F	100°C	373°K	212.0°F
-200	73	-328.0	200	473	392.0
-150	123	-238.0	300	573	572.0
-100	163	-148.0	400	673	752.0
-50	223	-58.0	500	773	932.0
-40	233	-40.0			
0	273	32.0	600	873	1112.0
20	293	68.0	700	973	1292.0
40	313	104.0	800	1073	1472.0
60	333	140.0	900	1173	1652.0
80	353	176.0	1000	1273	1832.0
100	373	212.0			

CENTIGRADE SCALE A scale for representing temperature, now officially called the Celsius scale. Also see *Celsius*.

CENTRIFUGAL COLLECTOR See Cyclone Collector.

CENTROSPHERE See Earth; Mantle.

CEQ Council on Environmental Quality.

CERENKOV RADIATION Light emitted by a charged particle when it is moving through matter with a speed that exceeds the phase velocity of light in the material.

CERMET A material composed of a mixture of a metal and a nonmetal, compacted and sintered together. Cermets are materials suited for high temperature use. These materials usually are made of TiC, Ni, Mo, and Cr_3C_2.

CETANE INDEX A number calculated to determine the suitability of a fuel for a Diesel engine, which parallels the Cetane Number.

The cetane index is calculated from the gravity and distillation point. Also see *Cetane Number.*

CETANE NUMBER An index for rating the anti-knock properties of fuels for Diesel engines in contrast to the octane number for fuels for the Otto cycle engine. The cetane number ranges from 35 to 60. Cetane number is obtained by using an engine to determine the knocking characteristics of the fuel, which is an expensive process. The calculated cetane index is used as a substitute for many purposes. Also see *Cetane Index; Diesel Fuel; Diesel Index; Octane Number.*

cf Cubic feet.

Cfd, cf/d Cubic feet per day.

CFFC Clean fuel from coal.

CFM, cfm Cubic feet per minute.

CFR Code of federal regulations.

CFR Cooperative fuels research (engine or procedure).

CGCC Coal gasification combined cycle.

CGS Centigrade-gram-second system of units. Also see *SI Units.*

C/H Carbon/hydrogen.

CHAIN REACTION A sequence of reactions each of which initiates the next member of the chain (Figure C-3). For example, a self-sustaining neutron chain reaction is a sequence of nuclear fissions, each fission producing neutrons that initiate the next fission in the sequence. Also called pile reaction.

CHAR After heating and distillation of a material, the solid remaining is known as char. The char can be further used as an energy source by using at a temperature higher than at which it was made.

CHARCOAL A substance consisting almost entirely of carbon formed by the thermal decomposition of wood. Charcoal may be used for fuel and as an absorbent to remove impurities from gas and liquids. Charcoal heated with steam to increase its absorbing power is called activated charcoal. Also see *Activated Carbon; Coke.*

CHARGE The quantity of material placed in a unit, either fuel or product being processed, for heating, processing, manufacturing, handling, etc. The charge may be expressed in terms of a batch, cycle of operation, unit of time (minute, hour, day).

Charge is used to represent the conversion of electric to

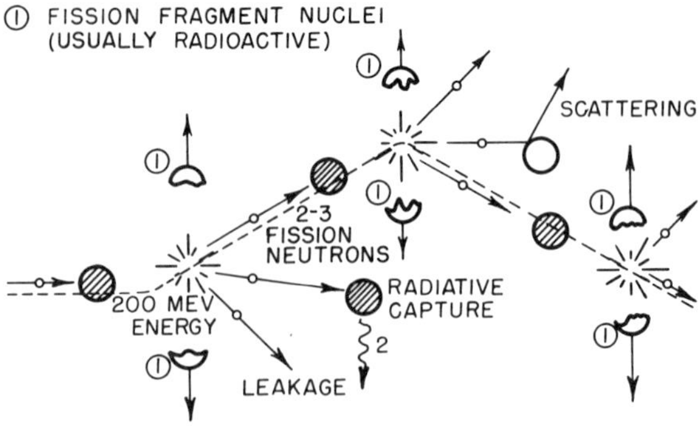

Fig. C-3 Chain reaction from fission *(Duderstadt, 1979)*.

chemical energy in a battery. The electric charge of a body represents the predominance of positive or negative charges, with protons being positive and electrons being negative.

The amount of explosive placed for blasting is called a charge.

CHEMICAL OXYGEN DEMAND (COD) A measure of the oxygen equivalent in a pollutant or effluent of that portion of organic matter that can be oxidized by a strong chemical oxidizing agent. Generally the BOD is 50-70% of the COD. The COD values can be obtained in two hours as compared to five days for the BOD. Also see *Biological Oxygen Demand*.

CHINA SYNDROME An expression used to describe the meltdown of a reactor core, resulting from lack of cooling, and the melting of the containment structure with flow of the molten material several feet into the ground toward China.

CIEC Conference on International Economic Cooperation.

CHLOROPHYLL The generic name for several green pigments present in chloroplasts of plants which are active in photosynthesis. Chlorophylls are photoreceptors of light energy and thus participate in the process of utilizing carbon dioxide and water to produce carbohydrates.

CIRCULATING FUEL REACTOR A reactor type which utilizes fuel in liquid form that circulates into and out of the reactor core. The molten salt reactor (MSR) concept studies in the past at Oak Ridge National Laboratory was a reactor of this type. One of the principal advantages of a circulating reactor would be the possibility of continuous fission product removal. At present there are no power reactors in the United States that use this concept.

CITIZEN'S ENERGY COUNCIL (CEC) An organization of community

leaders concerned with the hazards of nuclear energy. The address is Box 285, Allendale, NJ 07401.

CLADDING The covering on fuel elements in a nuclear reactor. In most present power reactors the cladding material is a metallic alloy using a low neutron absorption cross section material such as zirconium.

CLAMSHELL ALLIANCE An organization of people opposed to nuclear energy for power. Located at 62 Congress Street, Portsmouth, NH 03801.

CLATHRATE A solid-like substance composed of methane combined with water molecules in which the molecules of one are enclosed in the structure of another.

CLAUS PROCESS A process used to recover the sulfur in sour natural gas. The process consists of recovery of hydrogen sulfide followed by removal of the sulfur from the hydrogen sulfide. The process consists of an initial thermal stage where part of the hydrogen sulfide is burned. The following stages are catalytic in which the hydrogen sulfide in the gas and the sulfur dioxide formed from the combustion in the first stage react to form sulfur (S_2). Approximately 95% of the sulfur can be economically recovered from natural gas. Also see *Acid Gas; Frasch Process; Hydrogen Sulfide; Sour Gas*.

Clo A unit for designating thermal insulation as applied to clothing. It is the amount of insulation needed to permit maintenance of a skin temperature of a person at 33°C (92°F). The environmental conditions for the material are: Metabolic rate of 50 kcal/m_2-hr of body surface; 70°F ambient temperature; 50% or less relative humidity; and air flow of 20 fpm.

CLOSED CYCLE A system in which the fluid medium, such as in a refrigerator system, and in some gas turbine systems, is enclosed in the system and used over and over.

CLOSED-CYCLE REACTOR SYSTEM A reactor system in which the core coolant is recirculated. All current power reactors in the United States are of this type.

CLOSED-RING HYDROCARBONS Hydrocarbons with one or more closed rings (Figure C-4). These compounds are identified by the prefix cyclo, such as cyclopentane (C_5H_{10}) and cyclohexane (C_6H_{12}). Also see *Benzene*.

CLOUD CHAMBER A device used to make visible the paths of ionizing particles. The chamber has at least one wall that is transparent. The space inside the chamber is filled with a gas that becomes supersaturated with a condensible vapor, usually water, when the chamber is activated. The usual way to achieve the supersaturated condition is suddenly to increase the chamber volume by means of a movable chamber wall so as to reduce the temperature and pressure. During

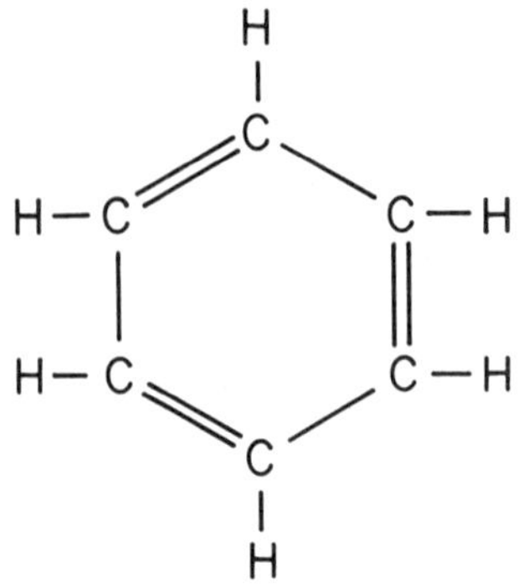

Fig. C-4 Closed ring hydrocarbon (benzene).

the supersaturated phase the ions created along the path of any ionizing particle act as nucleation sites for the supersaturated vapor. The droplets that appear along the path are visible and can be photographed through the transparent wall.

CLUB OF ROME A group of people concerned with the relationship of energy and environment and economies on a world basis, located at Via Giorgone 163, 00147, Rome, Italy. The Club issued a well-known study on Limits of Growth.

CNI Conservation needs inventory.

CNR Committee for Nuclear Responsibility.

COAL A solid fossil fuel, coal consists of carbon, hydrogen, oxygen, nitrogen, and moisture and other components, often undesirable from the aspect of combustion products, such as sulfur. The moisture content of coal varies from 2 to 20%. The higher the moisture content the more heat that is used to vaporize the moisture so that less heat is available for heating, unless heat is removed from the condensed vapor. After burning, 4 to 20% of the product remains as ash.

Anthracite, also known as hard coal, has a high content of carbon, 92 to 98%. Semi-anthracite has a carbon content of 86 to 92%. Anthracite is relatively smoke-free, ignites slowly, burns with a blue flame, and is non-coking. It has a net heating value of 5160 to 5590 kJ/kg (12,000 to 13,000 Btu/lb).

Bituminous, also known as soft coal, with 69 to 86% carbon, is used extensively for commercial purposes. It is generally more volatile than anthracite coal which makes it easier to ignite, will burn more rapidly, will coke, and the ashes will cake more easily. Bituminous coal has a net heating value of 5160 to 6020 kJ/kg (12,000 to 14,000 Btu/lb). Also see *Coal Classification; Coal Rank.*

COAL CAR The basic types of railroad cars used to haul coal are (1) open top hopper, (2) gondola, (3) gondola with rotary coupler, and (4) bottom unloading cars. Cars vary in capacity from 50 to 125 tons. Also see *Collier; Gondola Car.*

COAL CLASSIFICATION A method of categorizing coal of various types based on chemical composition, volatile content, calorific value, and agglomerating characteristics. Classification assists in specifying equipment for handling and burning coal. Anthracite coal is classified by fixed carbon. High volatile bituminous coal, bituminous coal, and lignite are classified by calorific values. The ASTM has established criteria for classification of coal. Also see *Coal; Coal Rank, Commercial Varieties (U.S.A.).*

COAL CLEANING The removal of foreign material from mined coal, usually by washing. The foreign material is usually heavier than the coal and the coal is floated from the washing and separating device. Mechanical dewatering will reduce coal to approximately 8% moisture; further removal, if required, is by thermal drying. Also see *Coal Drying.*

COAL DRYING The process of removing moisture from fine or pulverized coal, usually with one of three types of dryers: Semi-indirect-fired rotary dryer, cell-type dryer which uses hot flue gas from the stack, or steam heated dryers, in which the temperature inside the dryer is limited to about 121°C (250°F) to avoid over-heating of coal. To provide powdered coal for a boiler, the coal is first crushed between rolls to produce 2 to 2.54 cm (3/4 to 1 in) lumps, then dried, followed by pulverizing. Coal is dried to about 8% before pulverizing or milling. By drying, the capacity of the pulverizer is increased, dry coal packs less in the bin, lower stack losses result because a higher flame temperature is obtained, and the possibility of spontaneous combustion is reduced.

COAL GAS Also called town gas, coal gas is a mixture of carbon monoxide, ethylene, hydrogen, and methane, obtained by heating coal at a high temperature in the absence of air. A medium-Btu gas with 450-580 Btu/ft^3 (16.8-18.7 MJ/m^3). Also see *Medium-Btu Gas; Power Gas; Producer Gas.*

COAL GASIFICATION See Gasification.

COALIFICATION The natural processes by which plant materials are converted into coal. Peat is the initial stage, anthracite is the final stage. Also see *Carbonification; Carbonization.*

COAL LIQUEFACTION The process of producing liquid from coal (Figure C-5). Five coal liquefaction processes are being developed in the United States: H-coal (HRI), SRC II (DOE, EPRI, PAMCO, Gulf), SRC I (EPRI, Conoco, Kerr-McGee), and the Exxon Donor Solvent (EDS)

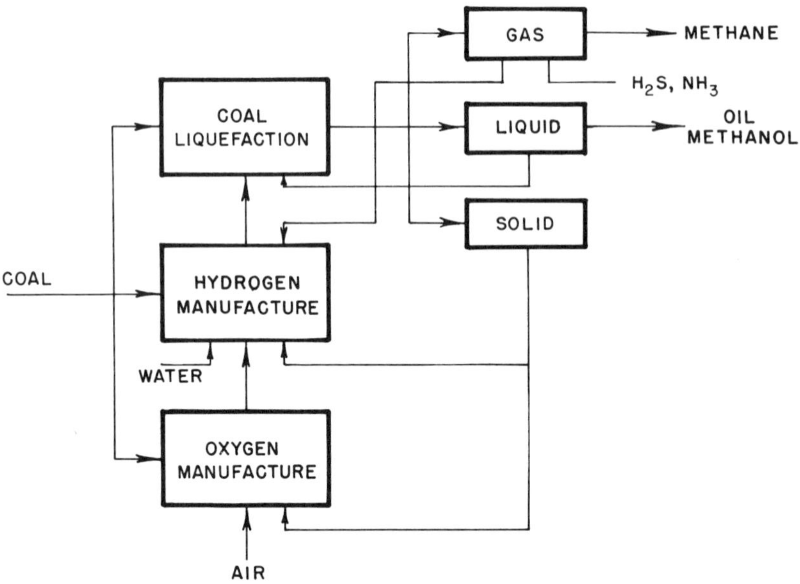

Fig. C-5 Coal liquefaction.

process. Also see *Consol Synthetic Fuel (CSF); Exxon Donor Solvent; H-Coal; Liquefaction (or Liquifaction)*.

COAL LIQUIDS Generally, the soluble products obtained when coal is heated, usually in the presence of a H-donor solvent. Also see *Coal Liquefaction; H-donor; Solvent Refined Coal (SRC)*.

COAL OIL A liquid consisting of several hydrocarbon components obtained by destructive distillation of bituminous coal. Also see *Coal Gas; Coal Tar; Kerosine*.

COAL RANK The classification of coal according to the evolution pattern from lignite to anthracite based on fixed carbon, volatile matter, moisture, oxygen, and heating values. There are 12 ranks of coal: meta-anthracite, anthracite, semi-anthracite, low-volatile bituminous, medium volatile bituminous, high volatile A bituminous, high volatile B bituminous, high volatile C bituminous, subbituminous A, subbituminous B, subbituminous C, lignite A, and lignite B. (See chart on the following page.)

COAL TAR One of the products of carbonization of coal, which with low temperature carbonization at 500-750°C (930-1380°F), also produces a coke. At high temperature a metallized coke is produced with coal tar. Also see *Coal Gas; Coal Oil; Metallurgical Coal*.

Classification of Coals by Rank (ASTM D 388)

Class	Group	Fixed carbon limits, %[1] Equal or greater than	Fixed carbon limits, %[1] Less than	Volatile matter limits, %[1] Greater than	Volatile matter limits, %[1] Equal or less than	Calorific value limits, Btu/lb[2] Equal or greater than	Calorific value limits, Btu/lb[2] Less than
I Anthracite	1. meta-Anthracite	98	-	-	2	-	-
	2. Anthracite	92	98	2	8	-	-
	3. Semianthracite	86	92	8	14	-	-
II Bituminous	1. Low volatile bituminous	78	86	14	22	-	-
	2. Medium volatile bituminous	69	78	22	31	-	-
	3. High volatile *A* bituminous	-	69	31	-	14,000	-
	4. High volatile *B* bituminous	-	-	-	-	13,000	14,000
	5. High volatile *C* bituminous	-	-	-	-	11,500	13,000
						10,500	11,500
III Subbituminous	1. Subbituminous *A*	-	-	-	-	10,500	11,500
	2. Subbituminous *B*	-	-	-	-	9,500	10,500
	3. Subbituminous *C*	-	-	-	-	8,300	9,500
IV Lignitic	1. Lignite *A*	-	-	-	-	6,300	8,300
	2. Lignite *B*	-	-	-	-	-	6,300

[1]dmmf - dry mineral-matter-free

[2]mmmf - moist mineral-moisture-free

Also see *Coal Classification*.

COAL TYPE A general term describing or classifying coal, more specifically identified by rank. Also used to describe coal on the basis of the plant material from which it evolved, such as woody plants, leafy plants, algae, etc. Also see *Coal Classification; Coal Rank.*

CO-CONVERSION or CO-COMBUSTION The burning of biomass blended with fossil fuel such as pelleted or briquetted sawdust with coal, which may have advantages of burning with a hotter flame, diluting of pollutants (sulfur) in coal, and providing a more complete combustion of biomass.

COD Chemical oxygen demand.

CODE OF FEDERAL REGULATIONS (CFR) The codified assembly of federal regulations passed by Congress and as developed through rulemaking procedure in the Federal Register. Most energy-related regulations are published under Title 10 of the Code of Federal Regulations, i.e., 10 CFR xxx.xx. Also see *Federal Energy Guidelines; Federal Register.*

COED Char Oil Energy Development.

COED PROCESS Char Oil Energy Development Process to convert coal to oil, gas, and char.

COFFERDAM A temporary structure, such as a dam, built to hold water away from a particular site while construction, maintenance, or repair can be accomplished. Also see *Caisson.*

COFIRING The adding of a second fuel to the primary fuel to improve combustion, such as gas to refuse.

COGAS PROCESS A method of treating char to produce gas. The char is a product of the COED Process.

COGEMA Compagnie Generale des Matieres Nucleaires.

COGENERATION The process in which fuel is used to product heat for a boiler-turbine or gas for a turbine, to drive a mechanical system or a generator to produce electricity, with the rejected heat used to produce process steam which can be used for additional power generation. There are two basic systems for cogeneration: Topping and bottoming. In the topping system electricity is produced first and then heat is produced. In the bottoming system heat is produced first which is then used to produce electricity. Also see *Bottoming Cycle; Topping Cycle; Tri-generation.*

COGP Commission on Government Procurement.

COG PLANTS Manufacturing plants which make coke, oil, and gas products from coal.

COKE A solid material obtained from carbonaceous materials such as wood, coal, and petroleum residues, which results from destructive distillation.

COKE A fuel consisting of the residue left after the volatile matter is removed from bituminous coal. Coke consists mainly of carbon (85-95%) which burns with a clean flame. Coke has a net heating value of 5160 to 6020 kJ/kg (12,000 to 14,000 Btu/lb). A coke burner as compared to a coal burner must have a larger combustion chamber because coke is more bulky than other fuels, otherwise more frequent firing is required. Coke can also be produced from the residue of the destructive distillation of petroleum. Coke is used as a fuel and as a source of carbon in the manufacture of steel and aluminum. Charcoal is a similar material produced from wood. Also see *Charcoal; Metallurgical Coal.*

COLLIER A ship used to carry coal. Also see *Barge.*

COLLOIDAL FUEL A fuel consisting of small particles of a solid fuel, such as coal, in fuel oil, and homogenized to stabilize the suspension. Used for stationary boilers.

COMBINED CYCLE In electrical power generation, a system in which a gas turbine drives a generator to produce electricity, and the hot exhaust (450°C) from the gas turbine is used to produce steam in a boiler, which is used to drive a steam turbine to drive a generator (conventional cycle) to produce electricity (Figure C-6). Thus, electricity is produced by a gas turbine and a steam turbine. A combined cycle might consist of a conventional cycle plus a topping cycle or bottoming cycle or both.

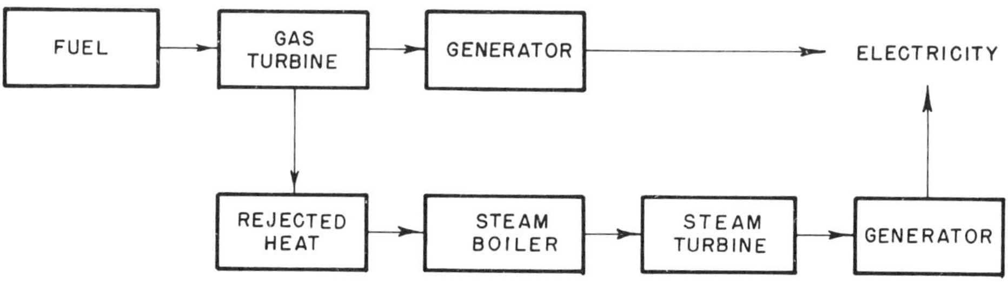

Fig. C-6 Combined cycle.

COMBUSTION The rapid oxidation of a fuel with the production of heat and light as in the burning of a fuel. Carbon, hydrogen, and sulfur are components of typical fuels which can be burned. With the complete oxidation of fuel in the presence of oxygen, carbon dioxide (CO_2), water (H_2O), and sulfur dioxide (SO_2) are produced.

Complete combustion is also known as stoichiometric combustion. The quantity of energy released is the higher heating value (HHV) or the lower heating value (LHV). The combustion of one pound of carbon produces 14,600 Btu (6277 kJ/kg); 1 lb of sulfur, 4000 Btu (1720 kJ/kg); and 1 lb of hydrogen, 62,000 Btu (2666 kJ/kg). The amount of air in excess of that theoretically needed to supply the oxygen for complete combustion for stoker-fed coal is 30 to 50%; fuel oil, 20 to 30%, and gas, 15 to 20%. Complete combustion occurs when all of the carbon is burned without an excess of air. Also see *Calorific Value; Carbon Dioxide; Carbon Monoxide; In Situ; Net Heating Value.*

Combustion Air Required for Complete Combustion at 1 Atm

	Lb air/lb of fuel	Major content of fuel, %	
		Carbon	Hydrogen
Coke	162.1	98	
Coal, hard	144.6	95	2.2
Coal, soft	143.4	87	5.0
Wood, dry	73.4	50	6.0
Natural gas	227.9	68	22.5

Conversion factor: *1 lb = 0.45356 kg.*

COMMERCIAL ENERGY The energy which is sold and purchased for commercial trade.

COMMERCIAL VARIETIES OF COAL These are defined by ASTM standards and are: Common banded coal, splint coal, cannel coal, and boghead coal.

COMMITTEE FOR NUCLEAR RESPONSIBILITY (CNR) A committee, not a membership organization, devoted to educating the public of the dangers of nuclear energy and supporting alternatives to nuclear energy. Address is P. O. Box 332, Yachats, OR 97498.

COMMITTEE TO END RADIOLOGICAL HAZARDS An organization devoted to ending use of atomic/nuclear energy for all uses. Located at 166 Second Ave., New York, NY 10003.

COMPRESSION REFRIGERATION A refrigeration system in which a fluid which is used in the liquid and gaseous states to transfer heat from an evaporator to a condenser. The system is a closed cycle. The gaseous fluid is compressed by a motor-driven compressor, thus the name compression refrigeration. Used in a reverse flow, the system becomes a heat pump.

CONCENTRATING COLLECTOR See Concentrating Ratio; Focusing Collector.

CONCENTRATING RATIO (CR) As applied to a concentrating or focusing solar energy collector, which is determined by dividing the opening or aperture through which the solar energy enters by the area of the absorbing surface.

Concentration ratios of solar energy collectors at equilibrium temperature, representative average values:

Type	Temperature	Concentration Ratio
Parabolic, following sun	3200°C (5800°F)	10,000
Heliostat solar furnace	2204°C (4000°F)	3,000
Conical concentrators	1649°C (3000°F)	1,000
Parabolic trough	1093°C (2000°F)	200
Solar cookers	760°C (1400°F)	100
Cylindrical trough	371°C (700°F)	10
Flat plate	93°C (200°F)	1

Based on H/P/A, Jan. 1981, p. 56.

CONDENSING SYSTEM In a thermal power system, a steam system where the rejected steam is condensed and returned to the boiler, and not used for other purposes such as heating or power.

CONDITIONING Using processes to change the physical properties of a product or material to improve the state of the product for further processing, storage, or packaging. Conditioning may be done by wetting with water or steam, by adding a chemical, by drying with forced unheated or heated air, or by equalizing and maintaining a uniform temperature of the product.

CONDUCTION (HEAT) The mode of heat transmission characterized by transfer of heat from one part of a solid to another part by molecular activity or in a fluid as distinguished from convection. Conduction is a mode of heat transfer distinguished from convection and radiation. The Fourier equation for thermal heat conduction under steady-state conditions is:

$$q = \frac{kA}{x}(t_2 - t_1) = (k/x)A\ (\Delta t)$$

where

- q = rate of heat transfer
- h = thermal conductivity of material
- x = thickness
- A = area perpendicular to direction of flow
- Δt = temperature difference along the path of heat transfer

Conductivity Coefficient, k, Conversion Factors

1 Btu-ft/hr ft^2°F = 12 Btu-in/hr ft^2°F
= 14.8816 gcal-cm/hr cm^2°C
= 4.1338 x 10^{-3} gcal-cm/s cm^2°C

Conductivity Coefficient, k, Conversion Factors (continued)	
1 Btu-ft/ft^2 hr °F	= 1.731 J/msK
	= 1.48816 kcal-m/m^2 hr °C
	= 1.731 x 10^{-3} kW/mK
	= 12 pcu-in/ft^2 hr °C
	= 1.731 W/mK
	= 17.31 x 10^{-3} W-cm/cm^2 °C
	= 43.931 W-in/in^2 °C

CONOZOIC See Era.

CONSERVATION OF ENERGY A principle or law which states that the total energy of a system or conversion remains constant. Energy is neither created or destroyed.

CONSOL SYNTHETIC FUEL (CSF) A process for converting coal to a liquid extract and a solid with the liquid extract hydrogenated to produce synthetic crude oil (Figure C-7). The process is used as the basis of the Cresap plant operation. Also see *Cresap Plant*.

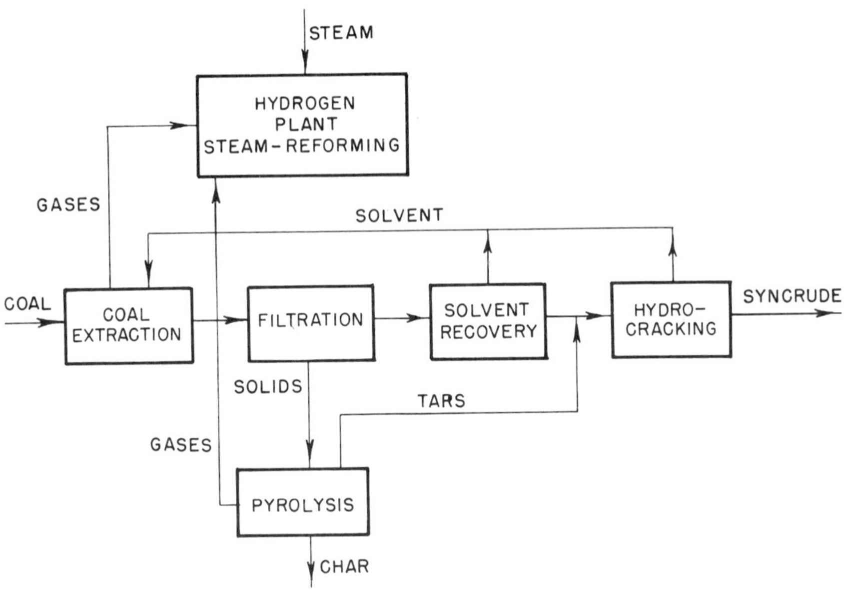

Fig. C-7 CSF process *(Kirk-Othmer, 1963-70)*.

CONSOLIDATED SYNTHETIC GAS (CSG) A process for gasification of coal, also known as a carbon dioxide acceptor process, for producing methane without supplying oxygen. Lime or calcined dolomite is used in the steam-carbon reaction, at about 870°C (1600°F) and 20 atm (300 psia).

CONSTANT A quantity which does not vary. A quantity which is the same for the conditions being considered in an equation, formula, or particular situation. Constant coefficients are used in equations. Constants, such as π, g, and e are used in many different calculations. Universal constants, such as those below, are of particular value:

Acceleration due to gravity, g = 9.81 m/s^2
Avogadro number, N_A = 602.25 x 10^{24}/kmol
Boltzmann constant, k = 13.805 x 10^{-24} J/K
Curie, Ci = 37.0 x 10^9 disintegrations/s
pi = 3.14159. . .
Planck constant, h = 662.6 x 10^{-36} Js
rad = 57.295779. . . deg
Stefan-Boltzmann constant, σ, = 56.696 x 10^{-9} W/m^2K^4
Universal gas constant, R = 8.3143 J/molK
Velocity of light, vacuum, c = 299.79 x 10^6 m/s

CONTAINMENT STRUCTURE A building surrounding a nuclear reactor pressure vessel and other equipment. The containment structure is designed to prevent the radioactive contents of the core from being released into the environment should an accident occur.

CONSOL SYNTHETIC FUEL PROCESS (CSF) A process of coal liquefaction to produce a liquid which can be used as a refinery feedstock. The process involves dissolving coal in a solvent at 400°C (752°F), with the solvent providing hydrogen. Also see *Coal Liquefaction; Liquefaction.*

CONTRACTED RESERVES The reserves of natural gas which have been committed for sale by contracts.

CONTRAROTATING BLADES As applied to a windmill for generating electricity, two sets of blades which operate in opposite directions. One set of blades drives the rotor and the other set of blades drives the field (normally static) in opposite directions giving twice the relative speed of a simple set of blades.

CONTROL RODS Rods that are inserted into a reactor core of a nuclear reactor to interfere with the self-sustaining chain reaction and put the system into a state of declining power level. Increasing operation can be restored by withdrawing the rods. The rods are usually made partially of a neutron-absorbing material such as boron, cadmium, or hafnium.

CONVECTION HEAT TRANSFER Transfer of heat, usually between a solid and a fluid, by moving masses of matter or by currents in gases and liquids caused by differences in density. Convection is the mechanism of heat transfer in which a combination of conduction within the fluid and energy transport is due to the fluid motion, produced by natural flow by density currents, known as natural or free convection; or by forced fluid flow, known as forced convection.

CONVECTION HEAT TRANSFER COEFFICIENT Represented by h, the transfer of heat by a surface film in convection heat transfer, Btu/hr $ft^2{}^\circ F$, W/m^2K, $W/m^2{}^\circ C$ or J/m^2sK. Also called the film coefficient of heat transfer. Used in the rate of heat transfer equation:

$$q = h\ A\ \Delta t$$

where

q = rate of heat transfer
h = convection heat transfer coefficient
A = surface area
Δt = temperature difference

Convection Heat Transfer Coefficients, h, Approximate

Condition	Btu/hr $ft^2{}^\circ F$	W/m^2K
Air, natural convection	1-5	6-30
Air, forced convection	2-100	10-600
Water, natural convection	10-80	50-450
Water, forced convection	20-3000	120-160,000
Water, boiling	500-10,000	2500-60,000
Steam condensing	1000-20,000	5500-110,000
Air inside tubes, 9.75 m/s (32 ft/sec)	8.0	45.4
Air perpendicular to tubes, 2.4 m/s (8 ft/sec)	7.5	42.6
Boiling water on vertical surface	1000	5678
Water inside tubes, 1.5 m/s (5 ft/sec)	1200	6815
Water perpendicular to tube	800	4542
Fluid paste	280	1590
Dry solid	2.5	14.2

Conversion factor: 1 Btu/hr $ft^2{}^\circ F$ = 5.678 W/m^2K = 4.886 kcal/hr $m^2{}^\circ C$.

CONVERSION OF UNITS The commonly used English units and the SI equivalents are:

Acceleration

$1\ ft/s^2 = 0.3048\ m/s^2$
$1\ g = 32\ ft/s^2 = 9.8067\ m/s^2$
$1\ in/s^2 = 0.0254\ m/s^2$
$1\ m/s^2 = 3.2808\ ft/s^2$

Area

1 acre = 0.4047 hectare = 4046.9 m^2
1 cm^2 = 0.155 in^2
1 ft^2 = 92.90 mm^2 = 0.0929 m^2
1 in^2 = 645.2 mm^2 = 6.4516 cm^2 = 0.00064516 m^2
1 km^2 = 0.3861 mi^2
1 m^2 = 10.76 ft^2 = 1550.15 in^2
1 mi^2 = 2.5899 km^2
1 yd^2 = 0.8361 m^2

Density

1 g/cm^3 = 0.0361 g/in^3 = 62.38 lb/ft^3 = 1 kg/m^3
1 k/l = 10.019 lb/gal Imp. = 8.3429 lb/gal U.S.
1 kg/m^3 = 0.0624 lb/ft^3 = 0.1603 oz/gal U.S.
= 0.13356 oz/gal Imp.
1 lb/ft^3 = 16.02 kg/m^3
1 lb/gal Imp. = 0.0998 kg/liter = 99.81 kg/m^3
1 lb/gal U.S. = 0.11986 kg/liter = 119.86 kg/m^3
1 lb/in^3 = 27.6799 g/cm^3 = 27679.9 kg/m^3
1 oz/ft^3 = 1.0012 kg/m^3
1 $slug/ft^3$ = 515.38 kg/m^3

Energy, Work, Heat (also see *Heat Transfer, Power)*

1 Btu = 1055.1 J = 1.0551 kJ = 252 gcal
= 0.252 kcal = 778.2 ft-lb
1 Btu/ft^2 = 2.712 $kcal/m^2$ = 11,356 J/m^2 = 11.356 kJ/m^2
1 Btu/ft^3 = 8.90 $kcal/m^3$ = 37.2826 kJ/m^3
1 Btu/lb = 0.566 kcal/kg = 2.324 kJ/kg = 2324 J/kg
1 Btu/lb °F = 4184 J/kgK
1 cal = 1 gcal = 4.184 J
1 Cal = 1 kcal = 4184 J
1 cal/g = 4184 J/kg = 1.8 Btu/lb
1 erg = 10^{-7} J
1 ft-lb = 0.1383 kg-m = 1.3558 J
1 ft pdl = 0.0421 J
1 $gcal/cm^2$ = 1 langley
1 J = 23.73 ft pdl = 0.7375 ft lb = 10^7 erg
= 0.94777 Btu
1 kcal = 3.9683 Btu = 4184 J
1 kcal/100 g = 41.84 kJ/kg = 18 Btu/lb
1 kcal/kg = 1.8 Btu/lb = 4.184 kJ/kg
1 $kcal/m^2$ = 0.3687 Btu/ft^2 = 4.183 kJ/m^2
1 $kcal/m^3$ = 0.1124 Btu/ft^3 = 4.183 kJ/m^3
1 kJ = 0.9478 Btu = 0.2388 kcal
1 kg-m = 7.233 ft-lb
1 kJ/kg = 0.4299 Btu/lb = 0.2388 kcal/kg
1 kWh = 3.600 MJ = 3415 Btu = 860.45 kcal
= 3,600,000 J
1 langley = 1 $gcal/cm^2$
1 therm = 10^5 Btu = 25.207 Mcal
1 W/m^2 = 0.8598 $kcal/m^2$ hr = 0.3170 Btu/ft^2 hr

Flow Rate

1 ft^3/min = 1.699 m^3/hr = 0.000472 m^3/s = 1 cfm
1 ft^3/s = 0.0283 m^3/s = 28.3198 liters/s = 1 cfs
1 gal/min = 0.063 liters/s
1 kg/s = 2.205 lb/s
1 lb/hr = 0.000126 kg/s
1 lb/s = 0.4537 kg/s = 27.223 kg/min
1 liter/s = 0.0353 ft^3/s = 0.220 gal Imp./s
= 0.2642 gal U.S./s = 15.85 gal U.S./min
1 m^3/hr = 0.589 ft^3/min = 3.666 gal Imp./min
= 4.4027 gal U.S./min
1 m^3/s = 35.3198 ft^3/s

Force

1 dyne = 1.020 x 10^{-3} g = 1.0 x 10^{-7} J/cm = 1.0 x 10^{-5} N = 7.233 x 10^{-5} poundals
1 kg-force = 9.8066 N = 2.205 lb-force
1 lb-force = 4.4482 N = 32.17 poundal
1 N = 10^5 dynes = 0.2248 lb-force = 0.10197 kg-force
1 poundal = 0.1381 N

Heat Transfer; Heat Flow

1 Btu/ft hr °F = 1 Btu-ft/ft^2 hr °F = 1.731 J/smK
= 1.731 W/mK = 1.488 kcal-m/m^2 hr °C
1 Btu/ft^2 = 2.712 kcal/m^2 = 11,356 J/m^2 = 11.356 kJ/m^2
1 Btu/ft^2 °F = 4.886 kcal/m^2 °C
1 Btu/ft^2 hr = 3.155 W/m^2 = 0.2712 gcal/cm^2 hr
= 2.712 kcal/m^2 hr
1 Btu/ft^2 hr °F = 5.678 J/sm^2K = 5.678 W/m^2K
= 4.882 kcal/m^2 hr °C
1 Btu-ft/ft^2 hr °F = 12 Btu-in/ft^2 hr °F
also see *1 Btu/ft hr °F*
1 Btu-in/ft^2 hr °F = 1/12 Btu-ft/ft^2 hr °F
1 Btu/lb °F = 4184 J/kgK
1 cal/cm sec °C = 418.4 J/smK = 418.4 W/mk
1 kcal/m^2 hr °C = 0.2049 Btu/ft^2 hr °F
1 W/mK = 0.5778 Btu/ft hr °F = 0.8598 kcal/m hr °C
= 1 W/m°C
1 W/m^2K = 0.1761 Btu/ft^2 hr °F = 0.8598 kcal/m^2 hr °C

Length

1 cm = 0.0328 ft = 0.3937 in = 393.7 mil
1 ft = 0.3048 m = 30.48 cm = 304.8 mm
1 in = 0.0254 m = 2.54 cm = 25.4 mm = 1000 mil
1 km = 0.6214 mile
1 m = 39.37 in = 3.281 ft = 1.094 yd
1 micron = 10^{-8} in = 10,000 Å
1 mil = 0.00254 cm = 25.4 microns = 0.001 in
1 mile = 1.609 km = 5280 ft = 1760 yd
1 mm = 0.03937 in
1 yd = 0.9144 m = 91.44 cm

Power

1 Boiler hp = 33,472 Btu/hr = 9.807 kW = 9807.3 W
1 Btu/hr = 0.293 J/s = 0.293 W = 0.298 x 10^{-4} boiler hp
1 ft-lb/s = 1.3558 W
1 hp = 0.746 kW = 745.7 W = 745.7 J/s = 550 ft-lb/s
= 33,000 ft lb/min
1 J/s = 1 W
1 kg-m/s = 9.8 W = 0.013 hp
1 kW = 1 kJ/s = 737.6 ft lb/s = 1.3410 hp
1 ton refrigeration = 3516.85 W = 3.51685 kW = 12,000 Btu/hr
1 W = 1 J/s = 0.7375 ft-lb/s = 3.412 Btu/hr
= 0.00134 hp = 0.102 kg-m/s = 44.24 ft-lb/min
= 0.737 ft-lb/s

Pressure

1 atm = 760 mm Hg = 33.9 ft H_2O = 14.696 lb/in^2 = 29.92 in Hg
= 1.0332 kg/cm^2 = 1.013 x 10^5 N/m^2 = 101.3 kN/m^2
= 1.013 x 10^5 Pa = 101.3 kPa = 1.013 bar = 760 torr
1 bar = 750 mm Hg = 14.5 lb/in^2 = 1 x 10^6 dynes/cm^2
= 10^5 N/m^2 = 10^5 Pa = 0.1 MPa
1 dyne/cm^2 = 1.0 x 10^{-6} bar
1 in H_2O = 25.40 kg/m^2 = 249.09 N/m^2 = 249.09 Pa
= 0.249 kPa = 0.0735 in Hg
1 in Hg = 3386.4 Pa = 3.3864 kPa = 3.3864 kN/m^2
= 13.595 in H_2O = 25.4 mm Hg
1 kg/cm^2 = 98066.5 Pa = 98.0665 Pa = 98.0665 kN/m^2
= 1000 g/cm^2 = 0.967 atm = 14.223 lb/in^2
1 kg/m^2 = 0.03937 in H_2O = 0.205 lb/ft^2
= 0.001422 lb/in^2 = 1 g/cm^2
1 kN/m^2 = 0.14504 lb/in^2 = 0.2953 in Hg
= 4.0161 in H_2O
1 lb/in^2 = 6894.8 Pa = 6.8948 kPa = 6.8948 kN/m^2
= 0.0703 kg/cm^2 = 0.068 atm = 2.036 in Hg
= 27.68 in H_2O
1 mm Hg = 0.01934 lb/in^2 = 133.32 Pa = 0.13332 kPa
= 133.32 N/m^2 = 0.0394 in Hg = 0.535 in H_2O
1 N/m^2 = 1 pascal = 1 Pa
1 psi = 1 lb/in^2 = 6894.76 N/m^2 = 6.8947 kN/m^2
= 2.036 in Hg = 27.68 in H_2O (see lb/in^2)
1 torr = 133.322 N/m^2 = 0.1333 kN/m^2 = 0.1333 kPa
= 1 mm Hg (abs) = 0.00136 kg/cm^2 = 0.0193 lb/in^2
= 0.001316 atm

Rotation; Movement; Vibrations

1 Hertz (Hz) = 1/sec (cycles, movements, revolutions, rotations, vibrations)
1 kHz = 1000 rps = 1000 cycles/s
1 rpm = 1/60 Hz = 1/60 rps
1 rps = 1 Hz = 60 rpm

Volume

1 ft^3 = 0.0283 m^3
1 gal U.S. = 0.003785 m^3 = 3.785 mm^3
1 m^3 = 1.6387 x 10^{-5} m^3 = 16.387 cm^3
1 liter = 1 dm^3 = 61.0246 in^3 = 0.26418 gal U.S.
= 0.21997 gal Imp. = 0.001 m^3
1 lb/ft^3 = 16.06 kg/m^3
1 lb/gal = 119.86 kg/m^3
1 m^3 = 1000 dm^2 = 35.31 ft^3
= 1.3075 yd^3 = 1 stere
1 oz/ft^3 = 1.0012 kg/m^3

CONUS Continental United States.

COOLING WATER Water from lakes, rivers, and streams used to cool electric power plants which is heated 10 to 15°F, during the process. Approximately 100 trillion gallons of water were used for this purpose in 1980 in the United States. A 1000 MW electric generation plant uses about 500,000 gpm of cooling water.

COOPERATIVE FUELS RESEARCH See Octane Number.

COP Coefficient of performance.

CORD A volume unit for cut wood, especially for fuel, equal to a pile of 4 ft x 4 ft x 8 ft or 128 ft^3 which will produce about 5 to 8 million Btu upon burning. The cord includes the air space between the wood pieces as well as the wood and will weigh about 1.2 tons. In contrast, board feet, as used for lumber, represents solid wood. The solid volume of a cord of wood is approximately 80 to 90 ft^3. Also see *Facecord; Short Cord; Thousand Feet Board Measure.*

CORRIM Committee on Renewable Resources for Industrial Materials (NRC, NAS).

COSMIC RADIATION High energy radiation arriving at the earth from outer space. The term also is applied to secondary radiation produced by the primary rays. The primary cosmic rays include various kinds of particles. They are predominantly positively charged.

COSMIC RAYS Radiation in space consisting of high energy particles and photons. The cosmic rays are continually bombarding the earth's atmosphere. Some rays penetrate to the earth's surface directly while others interact with atmospheric nuclei to produce a variety of secondary particles, including mesons, that travel to the surface.

COSMIC YEAR The time required for the sun to make one revolution around the center of the galaxy, approximately 200 million solar years.

COVALENT BOND A joining of atoms in which each atom contributes an electron to a shared pair constituting a chemical bond. Also see *Ionic Bond.*

CPA Canadian Petroleum Association.

CRACKING The process of breaking the molecular bonds of hydrocarbons to produce lighter factions such as gasoline and distillates. The cracking process results in products of lower molecular weight than the original petroleum. The major procedures for cracking are catalytic, hydro-, and thermal. Also see *Catalytic Cracking; Hydrocracking; Thermal Cracking*.

CRAP Calorific recovery anaerobic process.

CREOSOTE A mixture of oily substances, mostly cyclic, primarily phenols, resulting from distillation of coal or wood. Coal tar and wood tar are the liquids which are condensed from the distillation of coal and wood. Creosote is used as a wood preservative, antiseptic, or solvent.

CRESOL The oily substance distilled from heavy oil or coal tar consisting of a mixture of isomers of cresols. Cresols are poisonous and are used as a disinfectant.

CRESAP PLANT A CSF (Consol Synthetic Fuel) process for conversion of coal to a synthetic crude and a solid residue char byproduct, located in West Virginia. The process is carried out at 150 psi and 400°C (750°F). Also see *Consol Synthetic Fuel (CSF)*.

CRG Catalytic rich gas.

CRITICAL MASS The mass of nuclear fuel that must be included in the core of a nuclear reactor or other assemblage of materials to achieve a self-sustaining chain reaction. The number specified may be the mass of the fissile isotope or the mass of the material, e.g., enriched uranium oxide, considered to be the nuclear fuel.

CRITICAL PRESSURE The pressure at which a gas is liquefied when at its critical temperature.

CRITICAL TEMPERATURE The temperature above which a substance will exist only as a gas even in the presence of high pressures.

CRITICALITY A term describing the status of a neutron chain reaction. Criticality is achieved when the chain reaction is just self-sustaining so that the level of operation remains constant.

CROP COAL Coal located near or projecting from the surface, which, in comparison to underground coal, is usually of lesser quality.

CRT Cathode ray tube.

CRUST See Mantle.

CSF Consol Synthetic Fuel.

CSG Consolidated Synthetic Gas.

CSTR Continuous stirred tank reactor.

CTP Centro de Technologia Promon (Brazil).

CUBIC FOOT A volume dimension in English units, ft^3, equivalent to a cube with a length of one foot on a side. Cubic foot is used to represent the volume of gas, such as ft^3 in a container; flow of ft^3/min, or thermal energy, Btu/ft^3. One ft^3 equals 0.0283 m^3 or 28.316 liters.

CUBIC METER A volume dimension in meter units, m^3, equivalent to a cube with a length of one m on each side. Cubic meter equals 35.315 ft^3 or 219.97 Imperial gallons.

CULM Fine particles of coal or dust from anthracite coal mines.

CULTURAL ENERGY Refers to the energy which is used to support our way of life, or our cultural system.

CURIE A measure or unit of radioactivity, one curie equals 3.7×10^{10} atomic disintegrations per sec.

CVCS Chemical and volume control system.

CWA Clean Water Act.

CWT An abbreviation commonly used for one hundred weight, such as 20 cwt lb is 2000 lb or one ton. Also see *Ton*.

CYCLE A series of changes by a system or a component or the environment in which the change moves away from and back to the normal position, usually repeating the action and motion on a uniform repetitive basis, or frequency (such as for electricity, cycles per second; four seasons per year; etc.). The complete sequence of values of a periodic quantity which occurs during a period. A process for which the initial and final states are identical.

CYCLE OIL A liquid hydrocarbon which is a distillate of crude oil processing, which boils above gasoline, between the boiling point of the feedstock and gasoline.

CYCLING A periodic change of the controlled variable, such as the temperature of a heater or dryer.

CYCLING BURNER Type of operation wherein application of maximum heat is periodic such as cycling between high fire and the low fire; cycling between high fire and constant pilot; and cycling from high fire to "off", then restarted to high fire by constant or intermittent ignition.

CYCLONE COLLECTOR A conical device used to separate solids, normally in the range of 5 to 200 microns, from a swirling air stream. The solids are removed from the bottom of the cone and the air is exhausted from the center at the top. Proper design of the collector and air flow is required to remove the materials. The device is also used as a dust collector. Also called cyclone separator. Air with a high velocity enters the cone which has a much larger cross section than the entering duct. The air velocity decreases in the cone permitting the dropping of the solid. The following characteristics pertain to the cyclone separator: (1) The separation is a function of the difference in diameter and density of the product and air with the larger differences providing more efficient separation. (2) The velocity of the inlet is a major factor in controlling efficiency as related to the cyclone diameter. Velocities may approach 30.48 m/s (100 ft/sec). (3) A smaller diameter cyclone at a fixed pressure drop will have a higher efficiency than a larger diameter unit. (4) The diameter of the inlet is usually about one-fourth the diameter of the cyclone. (5) The collection efficiency can be increased by reducing the air outlet diameter, which will increase the pressure drop. (6) The pressure drop through the cyclone decreases as product, powder, or dust is incorporated in the air. (7) The pressure drop of the system decreases as the concentration increases for the same airflow in cubic feet per minute. Also called Centrifugal Collector; Dust Collector.

D

DAF, daf   Dry-ash free.

DAMP GAS   A gas deficient in oxygen from a mine.

DAMS   Earth or concrete structures used to hold water for reservoirs to provide a head for driving generators to produce electricity. Dams are made of earth or concrete. There are 65 hydroelectric plants with capacities over 1,000 MW in operation or under construction in the world. The highest dam in the world is the Nurck dam in USSR which is earth and is 317 m (1040 ft) high. Also see *Cofferdams; Grand Coulee; Low Head Hydro.*

DARIAZ TURBINE   A reaction hydraulic turbine in which water flows along the axis to turn a propeller. The blade surfaces on the propeller are oriented at an angle of about 45° with respect to the water flow and hence are called diagonal-flow turbines. Those diagonal-flow turbines with adjustable blades are called Dariaz turbines. Also see *Hydraulic Turbine; Kaplan Turbine.*

DARRIEUS ROTOR   A vertical axis windmill (aerogenerator) made of two or more thin strips for blades with the appearance of an egg beater. Wind coming from any direction can be used to drive a shaft providing mechanical work or electricity. The rotor is not self-starting. Also called wheel, turbine, or blade, and identified as VAWT, vertical axle wind turbine. At a given wind speed the Darrieus rotor operates at a higher speed than the Savonius rotor, the other major vertical axis windmill. Also see *Savonius Rotor; Wind Energy; Wind Turbine; Windmill.*

DAUGHTER ISOTOPE   The nuclide product remaining after decay by radiation of an atomic nucleus. A daughter is the nuclide formed in the decay of a parent. Following alpha particle emission, the daughter has an atomic number two less and an atomic mass four less than the parent. Following beta minus or electron emission, the daughter has one greater atomic number and the same atomic mass number as the parent. The daughter isotope is also called the decay product.

DEAD OIL A heavy oil, heavier than water, made from coal tar, and used for wood preservation, also called gallatin. Also refers to oil which is free of incorporated gas such as is present in live oil.

DEAD SPACE The volume of a vessel, tube, container, storage, dryer, etc., not actively involved or used in the containment of the active ingredient or product.

DEASPHALTING The process of removing asphalt components from a petroleum product, such as when using propane to dissolve the oil, with asphalt settling out.

DECANE A hydrocarbon, $C_{10}H_{22}$, of the paraffin series, which is a principal component of diesel fuel. Decane is followed by undecane, $C_{11}H_{24}$, in the paraffin series.

DECAY CHAIN A succession of radioactive decays in which each decay product is the parent nucleus for the next step in the chain. An example is the decay chain for radium:

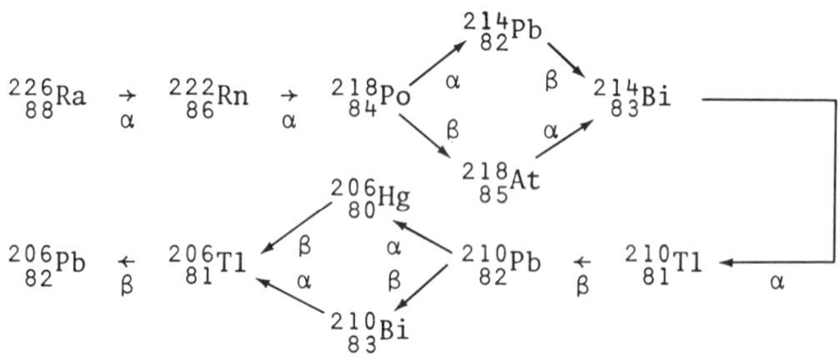

This is a branched chain as shown. Also see *Decay Product; Delayed Neutrons*.

DECAY CONSTANT or DECAY COEFFICIENT A number that indicates the rapidity of the decay of a radioactive material. The amount of radioactive material (n) present after time t is related to the initial amount (n_o) by $n = n_o e^{-\lambda t}$, where λ is the decay constant.

DECAY HEAT The nuclear energy, converted to heat, that results from the radioactive decay of fission products. Even after the chain reaction in a nuclear reactor is terminated, the reactor core continues to generate decay heat. Cooling must be provided to remove this heat because it cannot be turned off.

DECAY PRODUCT The nuclide that remains after a radioactive decay. For example, the decay product of the decay of ^{60}Co, a β-emitter, is ^{60}Ni.

DECOMMISSIONING Dismantling, entombing or in some other way deactivating a nuclear plant or other facility that will not be continued in use.

DECOMPOSITION The breakdown of minerals, for example, coal, as a result of natural processes such as weathering.

DEEP MINING The recovery of coal or ores at considerable depth in the earth, usually at least 1000 meters. The deepest coal mine in the world is in Belgium at 1365 m (4462 ft) with a rock temperature of 52°C (126°F); the deepest gold mine is in South Africa at 3120 m (10,200 ft).

DEGASIFICATION The process of removing gas and volatiles from oil in a reservoir by use of separators on the surface. Separation is done by pressure differences as well as with a stripping unit. A separator is a device incorporating a momentum absorber to remove the liquid droplets from the gas stream, with gas removed at the top and liquid removed at the bottom of the unit. Also see *Gas/Oil Ratio*.

DEGASSING The process of removing gas from oil at the well. If there is a large amount of gas removed at the well which cannot be sold the gas is forced back into wells or reservoirs. Also see *Dry Gas; Gas/Oil Ratio*.

DEGRADATION OF ENERGY The conversion of energy to a form less readily available to do work. Generally electrical and mechanical energy are fully available to do work, but thermal energy becomes less available as the temperature at which it can be delivered is lowered.

DEGREE-DAY A measure of the departure of the mean daily outdoor dry bulb temperature from a standard. Standards commonly used are 18.3°C (65°F) for heating or cooling, 4.4°C (40°F) for crop production, and 0°C (32°F) for contractors. The degree-day is calculated by summing the difference between the mean daily temperature and the standard for a period of time. If the maximum daily temperature is 32.2°C (90°F) and the minimum is 15.6°C (60°F), the mean is 23.9°C (75°F) and the degree-day is +10°F for that day, based on the 18.3°C (65°F) standard. Should the same mean persist for 30 days, the degree-day would be +300 (based on °F) for that month. Air drying would normally occur more rapidly, or less fuel would be required for heated air drying, during a time of greater degree-days (Figure D-1).

DEHUMIDIFY The process of decreasing the water vapor or the humidity of a body of air or gas such as through condensation on a cold surface, diffusion of vapor to an adsorbing agent, chemical reaction, and heating.

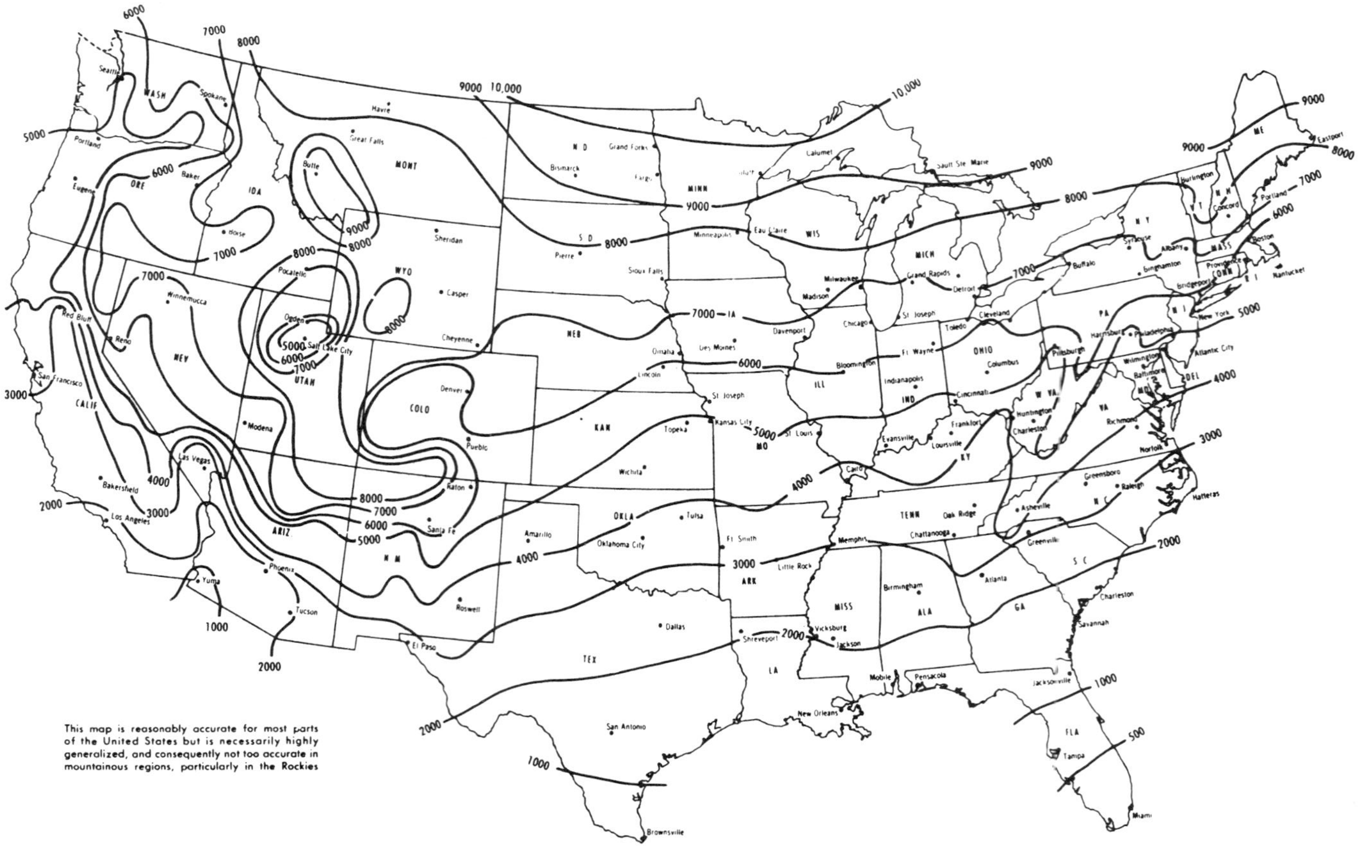

Fig. D-1   Degree-day map of the United States. *Reprinted with permission from Handbook of Air Conditioning, Heating and Ventilating, Strock and Koral, Industrial Press, New York, 1965.*

DEHYDRATION The process of removing water, usually to preserve the properties of substances. Water of crystallization in many salts is eliminated by heat. Alcohols give up water when heated to 400 to 500°C (752 to 932°F) in the presence of alumina. As distinguished from sun drying, dehydration involves controlled artificial heating. Also see *Coal Drying*.

DEHYDROGENATION The process of removing hydrogen which is associated with metallic reforming, a process of converting low octane to high octane fuels.

DEHYDROCYCLIZATION The process of treating certain paraffins to produce aromatic hydrocarbons.

DEIONIZATION A process that eliminates ions or charged particles.

DELAYED NEUTRON FRACTION The fraction of neutrons in a nuclear reactor that are not produced directly during the fission process but that come from radioactive decay or neutron-emitting nuclides in the fission product decay chains. The delayed neutron emitters assist in controlling the power level of a nuclear reactor.

DELAYED NEUTRONS Neutrons that are produced during the decays of certain radioactive nuclides. All of the delayed neutron emitters occur in the decay chains of fission products. Also see *Decay Chains*.

DEMAND The quantity of power required by consumers. The demand may be expressed in terms of horsepower or watts. Demand meters may be used to measure the maximum demand during a certain period of time. Also see *Demand Charge*.

DEMAND CHARGE A utility charge to a customer based on maximum potential demand for power/energy rather than the actual power/energy demanded.

DEMAND FACTOR The ratio of the maximum demand for power to the connected load, particularly applicable to electric power systems.

DENATURANT A substance added to alcohol so that it cannot be safely consumed. For example, an additive to alcohol to be utilized as a solvent. Also, a substance added to certain fissionable materials to make them unsuitable for weapons use, as by adding ^{238}U to ^{235}U. Also see *Alcohol*.

DENATURATION An irreversible change of property in a protein produced by excessive heating as in cooking, drying, or dehydration, which results in decreased solubility, less emulsifying ability, less water-holding capacity, and reduced rehydratability. Also see *Protein*.

DENSITY, AIR The weight per unit volume of air, given in pounds per cubic foot, grams per liter, or kilograms per cubic meter at a specified temperature. (See Table D-1.)

TABLE D-1
Density of Air and Vapor at 1 atm

Temperature		Density, lb/ft^3 Dry air	Saturated air
0°F	-17.8°C	0.0864	0.0863
10	-12.2	0.0845	0.0843
20	-6.7	0.828	0.0823
32	0	0.0807	0.0802
40	4.4	0.0794	0.0788
50	10.0	0.0779	0.0769
60	15.6	0.0764	0.0750
70	21.1	0.0751	0.0731
80	26.7	0.0735	0.0710
90	32.2	0.0722	0.0688
100	37.8	0.0709	0.0663
150	65.6	0.0651	0.0486
200	93.3	0.0601	0.0130

1 lb/ft^3 = 16.018 kg/m^3
1 ft^3/lb = 0.0624 m^3/kg

DENSITY, GAS The weight per unit volume of gas, usually expressed in terms of pounds per cubic foot or grams per liter or kilograms per cubic meter, at 0°C (32°F), at atmospheric pressure, unless otherwise specified (Table D-2). Specific gravity values of gases are based on a particular gas, such as air or oxygen, assigning SG = 1 for the reference gas.

TABLE D-2
Density of Gases at 1 atm and 0°C

Gas	g/L	lb/ft^3	Specific gravity based on air
Air	1.2929	0.0807	1.00
Carbon dioxide	1.9769	0.1234	1.529
Carbon monoxide	1.2504	0.07806	0.9671
Hydrogen	0.0898	0.00561	0.0695
Nitrogen	1.2505	0.07807	0.9672
Oxygen	1.429	0.08921	1.1052

DENSITY, WATER The weight per unit volume of water in pounds per cubic foot or grams per cubic centimeter or kilograms per cubic meter (Table D-3). The temperature of maximum density for pure water is at 4°C (39.2°F), the density is 1 gram/cm^3. The density decreases as the temperature increases above 4°C.

TABLE D-3
Density of Water

Water Temperature		Density gm/cm^3	lb/ft^3
39.2°F	4°C	1.00	62.43
50	10	0.99973	62.41
122	50	0.98807	61.68
212	100	0.95838	59.83
392	200	0.8628	53.86

Conversion factor: *1 lb/ft^3 = 16.018 kg/m^3*

DEPARTMENT OF ENERGY (DOE) A major cabinet-level department in the United States government, established in 1977 as a successor to the Energy Research and Development Administration (ERDA) which also incorporates the former Federal Power Commission (FPC) and the former Federal Energy Administration (FEA). Many energy-related activities of several of the other departments were assigned to DOE. The DOE is charged to provide unified leadership and to develop and carry out the nation's energy programs. Most of the federal departments—agriculture, commerce, interior, defense—have an office, agency, or directorate for energy. Also see *Economic Regulatory Administration; Energy Information Administration; Federal Energy Regulatory Commision; Nuclear Regulatory Commission.*

DEPA Defense Electric Power Administration.

DEPOLYMERIZATION A change of long or large molecules into short or small molecules such as aromatics, as with coal, usually accompanied by substitution of hydrogen or oxygen under heating processes. The opposite of polymerization.

DERIVED FUEL A fuel which is manufactured from an energy source, such as coke from coal, charcoal from wood, gasoline from petroleum, etc.

DESICCANT A chemical or material which takes on moisture to dry another material, usually a gas, or by drying a gas which has absorbed (adsorbed) moisture from a liquid or solid.

Silica gel and activated ammonia are adsorption desiccants which dry by surface attraction. Magnesium perchlorate and calcium sulfate are chemical desiccants, which dry by chemical reaction with water to form hydrates. Magnesium perchlorate is one of the best adsorbers and will adsorb water to about 50% of its dry weight.

DESORPTION The process of removing moisture from a solid. Desorption is the reverse process of sorption or absorption.

DESTRUCTIVE DISTILLATION A process in which heat is applied to solids, such as coal, wood, biomass, and solid wastes, to cause decomposition usually at temperatures of 500° to 1000°C (932° to 1832°F) to produce gases, volatile fluids (which might be condensed to liquids), coal, tar, char, and coke. Also see *Condensation; Distillation; Gasification.*

DESULFURIZATION The process of removing sulfur from a mineral such as coal, petroleum, or gas. A commonly used process to remove sulfur from gas is the Claus Process. Removal of sulfur from solids and liquids may consist of heating followed by condensation of the sulfur gas produced. Also see *Claus Process; Fluid Desulfurization; Solvent Extraction.*

DEUTERIUM A stable isotope of hydrogen, often called heavy hydrogen, which has double the mass of hydrogen, an atomic weight of 2, represented by D. Deuterium is a potential fuel for thermonuclear fusion power reactors. At room temperature it is an odorless gas with a density of 0.18 grams per liter (two times as heavy as hydrogen), and a boiling point of -249.5°C (-417°F). Also see *Deuteron; Heavy Water.*

DEUTERON A nucleus containing one proton and one neutron. A deuteron plus a surrounding electron constitute a deuterium atom. Deuterium is a form of heavy hydrogen. Also see *Deuterium.*

DEWATERING COAL The removal of water following washing of coal. Vibrating screens are used to remove free water, with larger particles over 1 cm (1/2 in) diameter dried to about 4%, and smaller particles dried to about 10%. Additional water removal can be by centrifugal action as well as heated air drying. Also see *Coal Drying.*

DIALYSIS The process of separating small molecules and ions from large molecules and colloidal materials using semipermeable membranes. Also see *Electrodialysis.*

DIESEL ENGINE An internal combustion engine in which the fuel-air mixture is ignited by the heat of compression of the gas in a cylinder (Figure D-2). The compression ratio will usually be between 12 and 20. The fuel for a diesel engine is of a lower grade and usually less expensive. An injector is used to feed the fuel into the combustion chamber under pressure. Diesel engines have a higher thermal efficiency than conventional spark ignition engines. Also see *Cetane Number; Diesel Fuel.*

DIESEL FUEL; DIESEL OIL A fraction produced by distillation in petroleum processing which is used for fuel in diesel engines. Diesel fuel is separated in the medium range of distillates, with a boiling point of 250 to 350°C (482 to 662°F). As compared to gasoline, diesel fuel in a diesel engine provides a greater efficiency of conversion because of the higher temperature of combustion and recovery of heat. Also see *Cetane Index; Cetane Number; Decane; Diesel Engine; Diesel Index; Octane Number.*

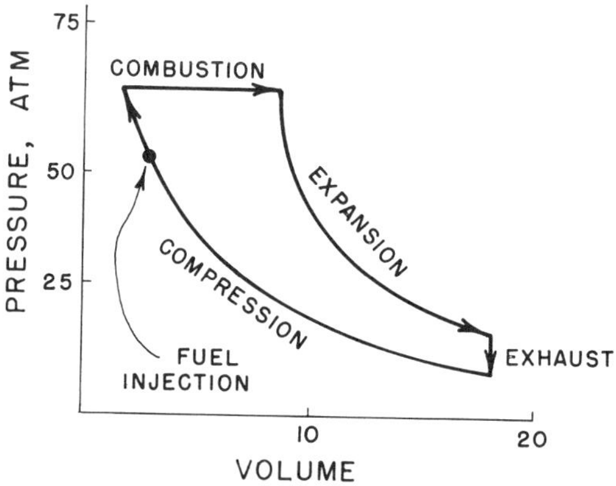

Fig. D-2 Diesel cycle.

DIESEL INDEX A number which is calculated from the gravity and aniline point of a fuel which represents the ability to be used in a diesel engine without knocking, similar to the Cetane Index. Also see *Cetane Index; Cetane Number; Diesel Fuel; Octane Number*.

DIESOHOL A blend of ethanol and diesel fuel, conventionally as 10% anhydrous ethanol and 90% No. 2 diesel fuel. Consideration is given to injecting diesel fuel with ethanol being vaporized into the air intake and to mixing the fuels before injecting. Characteristics of a 10/90 diesohol are: Cetane number, 45; gross heat, 37.5 MJ/L; air/fuel ratio, 14.6; and viscosity, 2.1 mm^2/s. Also see *Gasohol*. *(Wrage and Goering, 1980.)*

DIMENSIONS As applied to a physical quantity, the power or exponent of fundamental units of measure. The fundamental units are length, ℓ; time, t; and mass, m. Thus, area is ℓ^2; volume, ℓ^3; velocity, ℓ/t, in which dimensions for ℓ are 2 for area, 3 for volume, and 1 along with t which is -1 for velocity. In SI units the fundamental units are length, meter, m; time, seconds, s; and mass, kilogram, kg. Also see *Fundamental Units*.

DIRECT-FIRED DRYER A type of dryer in which the products of combustion for heating come into direct contact with the product being dried. Products which are not unfavorably affected by the products of combustion may be dried with a direct-fired unit. Direct-fired units have a thermal efficiency of 45 to 60%. Also see *Indirect-Fired*.

DIRECT MODE A means for providing thermal energy for the retorting process. The carbonaceous residue of the retorted shale and the low-Btu gas from the retort are burned in the combustion zone.

DIRECTLY IONIZING PARTICLES Charged particles which have enough energy to produce ionization by collision.

DISASTER See Major Mine Disaster.

DISTILLATE; DISTILLATE FUEL The fuels obtained by the fractional distillation of crude oil. From an input of crude oil the following are produced: Condensate; light fractions at 30 to 250°C (86 to 482°F) which includes gasoline; medium or middle fraction, at 200 to 350°C (392 to 662°F), sometimes called gas oil and includes light fuel oil; and the heavy oils which have a high boiling point; and the residuals. Also see *Distillation; Gas Oil*.

DISTILLATION Evaporation of a liquid followed by condensation of the gases. Condensation can be done at different temperatures so as to separate the constituents of the original liquid.

The range of fuels obtained from the distillation of crude oil is as follows:

Gases	less than 0°C (32°F)
LPG	between -30 and +30°C (-22 and 86°F)
Gasoline	30 to 200°C (86 to 392°F)
Kerosine	150 to 250°C (302 to 482°F)
Gas oils	200 to 350°C (392 to 662°F)

DISTILLERS GRAIN See Brewers Dried Grain; Stillage.

DISTRICT HEATING A method of providing space heating within a district of a city by using the steam or hot water from a central heating plant, power plant or industrial "waste" heat source.

dmmf Dry, mineral matter free.

DNB Departure from nucleate boiling.

DOE Department of Energy (U.S.A.).

DOPE A material, considered an impurity, added to certain metals to make semiconductors, such as for solar cells. Sawdust, used as an absorbent in making dynamite, is called a dope. Additives for lubricants and fuels are sometimes called dopes. The liquid put on pipe threads to provide a secure joint may be called a dope. A thick grease used on casing threads and drill rod may be called a dope.

DOSE, ABSORBED The energy imparted to matter in a volume element by ionizing radiation divided by the mass of irradiated material in that volume element. Often called dose. Also see *Rad*.

DOSE DISTRIBUTION FACTOR A numerical factor used in computing dose equivalent to account for the non-use, from distribution of internally deposited radio nuclides. Also see *Dose Equivalent, rem.*

DOSE EQUIVALENT The product of absorbed dose, quality factor, dose distribution factor, and other modifying factors necessary to express on a common scale, for all ionizing radiations, the irradiation (biological damage) incurred by exposed persons. The unit of dose equivalent is the rem.

DOT Department of Transportation (U.S.A.).

DOUBLING TIME In one usage the period of time required for the fuel inventory in a breeder reactor to double. In another usage it is the period of time during which the power level in a nuclear reactor doubles.

d•RDF Densified refuse-derived fuel.

DRILLING The forming of a hole in the earth to recover or obtain water, hydrocarbons, or steam, or to obtain information on the formation of the earth and its contents. The various types of drilling are: Cable tool, diamond, percussion, and rotary. Also see *Quebracks*.

DRIVER FUEL The fuel that provides the critical mass in a nuclear reactor. The driver fuel is distinguished from fuel that builds up in the blanket section of a reactor.

DRY BASIS A method of representing the moisture content of a product, either as a decimal or percent, in which the weight of water, W_w, in a given volume of product is divided by the weight of dry matter, W_d, dry basis. The dry basis of moisture content is used mainly by researchers, in scientific writings, and in equations dealing with moisture content variations. The moisture content, dry basis, percent $= W_w/W_d \times 100\%$. The dry basis moisture content is usually more acceptable than the wet basis moisture content when making calculations because the dry weight remains constant whereas the wet weight used in the wet basis representation changes during drying. The value of the moisture content on dry basis is larger in magnitude than the wet basis, for the same water content. The relationship between moisture content, dry basis, percent, MC_{db}, and wet basis, percent, MC_{wb}, is given by:

$$MC_{db} = \frac{MC_{wb}}{100 - MC_{wb}}$$

DRY GAS A gas produced without or with only a small amount of liquids, either from a dry gas field, or the name given to the gas after liquids are removed. Dry gas is primarily methane with some ethane, but excluding significant amounts of propane and butane and the high boiling point fractions of hydrocarbons. Also see *Degassing*.

DRY STEAM Water in the vapor form in which all of the water has been evaporated; in contrast to wet steam, in which water is contained in the liquid form. Steam in which all of the water has been changed to a vapor, but to which no additional heat is added is saturated steam. Also see *Superheated Steam; Wet Steam.*

DUAL-CYCLE REACTOR SYSTEM A nuclear reactor system that has both primary and secondary cooling circuits. The coolant in the primary circuit transfers heat from the core to a heat exchanger where the heat passes into a secondary circuit. The coolant in the secondary circuit is converted to steam in the heat exchanger. The steam is used to operate a turbogenerator before it is condensed and returned to the heat exchanger.

dwt, DWT Dry weight ton(s).

DYNE A fundamental unit of force in the c-g-s system which provides an acceleration of 1 cm/sec^2 to 1 gram. 980.7 dynes = 1 gram-force. The fundamental unit of force in the SI units is the newton, N, which is 10^5 dynes.

E

e The base of the natural logarithm which has a numerical value of 2.71828. e is the sum of the infinite series

$$1 + 1/1! + 1/2! + 1/3! \ . \ . \ . + 1/n!$$

Often written as exp preceding a part of an equation which should be raised to the power of e; e.g.,

$$A + B \exp (C + D) \text{ is } A + B\, e^{C + D}.$$

EARTH The cross section of the earth consists of the outer layer or crust, lithosphere; the under layer, beginning about 10 km below the surface, the mantle, or asthenosphere; and the large core in the center, centrosphere. The earth is considered to be approximately 4.5 billion years old. Also see *Magma; Mangle; Mohole.*

EARTH RESOURCES TECHNOLOGY SATELLITE (ERTS) A man-made satellite which continuously orbits the earth obtaining spectral data related to resources of the earth and transmits these data to the earth. Satellites are usually at least 200 mi above the earth. The program was formerly called EROS, Earth Resource Orbitting Satellite.

EASTERN OVERTHRUST BELT See Overthrust Belts.

EBR Experimental breeder reactor.

EBR-1 The experimental breeder reactor No. 1, a historic reactor, formerly located in Idaho. It was the first reactor to produce electricity and was used for many years as a test facility for fast reactor components. It has now been decommissioned.

EBTR Emergency building temperature restrictions.

EBULLATING BED A boiling bed reactor used to heat and provide chemical reactions for hydrocarbons or carbohydrates for gasification or liquefaction processes. An ebullating bed reactor is used for hydrocracking and hydrodesulfurization processes.

ECCS Emergency core cooling system.

ECNP Environmental Coalition on Nuclear Power.

ECOLOGY The field of study of the interrelationships between organisms and their environment, also known as environmental biology. An ecologist is one who works in the field of ecology. Also see *Environment*.

ECONOMIC COAL RESERVES The coal resources which can be economically removed at the time of determination or as projected to be economical at some specified time. Thus, the economic coal reserves vary with economics.

ECONOMIC REGULATORY ADMINISTRATION (ERA) A unit of the Department of Energy which administers regulatory programs except those which are the responsibility of FERC. The ERA is a successor of the Federal Energy Administration (FEA). The Administration has been responsible for oil pricing, allocation, and import programs; natural gas emergency allocation and planning; and natural gas import/export controls.

ECOSPHERE The biosphere along with the ecological factors which affect the organism. Also see *Biosphere; Ecology; Environment*.

ECOSYSTEM A natural interdependent, self-sustaining system of living organisms along with the environment in which they live. Also see *Biosphere; Ecology; Ecosphere; Environment*.

EDC Export Development Corporation (Canada).

EDF, EdF Electricite de France.

EDP Environmental Development Plan.

EDR Equivalent direct radiation.

EDS Exxon Donor Solvent process.

EEC European Economic Community.

EEI Edison Electric Institute.

EEPA Emergency Electric Power Administration.

EER Energy efficiency ratio.

EFFECTIVE HALF-LIFE (nuclear medicine) The time required for the activity of a radioisotope absorbed by the body to decrease by a factor of two. The decrease occurs because of the physical decay of the radioisotope and the gradual elimination of the radioisotope from the body. The physical (P), biological (B), and effective (E) half-lives are related by

$$\frac{1}{E} = \frac{1}{P} + \frac{1}{B}$$

EFFECTIVENESS For a heat pump or refrigerator the ratio of the energy transferred into a given space, such as a room, to the energy expended to cause the transfer. The same as the Coefficient of Performance. Also see *Coefficient of Performance; Energy Efficiency Ratio.*

EFFICIENCY Broadly defined, efficiency is the ratio of useful energy output to total energy input, usually expressed as a percent. The efficiency can be applied to an element, process, or system. In a system of several elements, the overall efficiency is the product of the efficiency of each element (Table E-1).

TABLE E-1
Efficiency of Power Plant and Comparative Costs

	Plant efficiency, %	Cost of electricity, mills/kWh
Gas turbine/steam turbine, combined	50	67-87
MHD	49	33-39
IGCC/3000°F	42	35-41
Advanced fixed bed/air (3000°F)	42	35-41
Advanced fixed bed/air (2600°F)	41	40-46
IGCC/2400°F	40	40-45
Pressurized fluidized bed (PFB)	40	39-45
Atmospheric fluidized bed (AFB)	36	39-45
Conventional boiler	32	47-54

IGCC integrated gasification combined cycle. *(Corman, 1979.)*

Also see *Carnot; Energy Conversion; Energy Efficiency for Illumination; Energy Efficiency Ratio; Food System; Freight Efficiency; Passenger Efficiency.*

EFFLUENT A general term for a gas or liquid emitted from a source, such as from a waste disposal system, power plant, factory, farm, or home. Some effluents contribute to the environmental pollution of the air, water, or land. Also see *Thermal Pollution*.

EHV Extra high voltage. See Voltage Classification.

EIA Environmental impact assessment.

EIA Energy Information Administration.

EIS Environmental impact statement.

ELECTRICITY GENERATION The production of electricity using primary fuels such as coal, oil, wood, gas, hydroelectric, and nuclear. Most electricity is generated using steam produced by a heat source utilizing fossil fuels or nuclear energy. Gas and oil can be used for combustion turbine drives so that the steam heating cycle is avoided. Hydroelectric generation is a process for changing the gravitational potential energy of water to mechanical energy of a rotating shaft and then to electrical energy in an electric generator. Present day operations utilize about three tons of oil equivalent (TOE) (toe) to produce one toe of electricity, or a conversion efficiency of 33 to 35%.

ELECTRIC POWER, VOLTAGE See Voltage Classification.

ELECTROCHEMICAL CELL A battery which produces a current flow as a result of a chemical reaction.

ELECTROCHEMICAL SERIES See Electromotive Force.

ELECTRODIALYSIS The use of electric current applied to electrodes on the two sides of a membrane to assist in dialysis. The electric current causes the migration and separation of saline cations and anions through a solution and a set of permeable membranes. Also see *Dialysis*.

ELECTROLYTE A chemical compound which conducts electricity as a result of the presence of ions or charged particles. Acids, bases, and salts are common electrolytes.

ELECTROMOTIVE FORCE (EMF) The voltage produced at electrodes of an ionized solution or at terminals of a generator. Metallic elements and alloys can be placed in order of most active to least active, providing an EMF series, also known as an electrochemical series. The electromotive series of some elements from most to least active include:

Lithium	Manganese	Nickel	Bismuth
Potassium	Zinc	Tin	Mercury
Calcium	Chromium	Lead	Silver
Sodium	Iron	Hydrogen	Platinum
Magnesium	Cadmium	Copper	Gold
Aluminum	Cobalt	Arsenic, Antimony	

ELECTROMAGNETIC RADIATION The result of an electric or magnetic field which produces various effects at different wavelengths, extending from gamma rays (short) to radio waves (long). The range of radiation is represented by the electromagnetic spectrum, Table E-2.

TABLE E-2
Electromagnetic Spectrum

Type of Wave	Wavelength
Radio	above 1000 km to 10 m
Short wave	10 m to 1 m
Ultra short wave	1 m to 0.1 cm
Infrared	0.030 to 0.000076 cm
Ultraviolet	0.000040 to 0.0000013 cm
X-rays	10^{-6} to 10^{-4} cm
Gamma rays	10^{-8} to 5×10^{-11} cm

Also see *Gamma Rays; Infrared; Ultraviolet; X-rays.*

ELECTRON CAPTURE A nuclear reaction in which one of the electrons orbiting a nucleus is captured by that nucleus. As a result of the reaction the nuclear mass number does not change but the nuclear charge number is reduced by one unit. The electron usually is captured from the K-shell but occasionally comes from one of the other shells.

ELECTRON VOLT The energy gained by one electron moving through a potential, or voltage, of difference of one volt. The value is 1.6×10^{-19} J. A MeV is a million or mega electron volts. A BeV is a billion electron volts.

ELECTROREFINING The process of separating impurities or unwanted materials from a desired metal by electrolysis.

ELECTROSTATIC PRECIPITATOR (ESP) A device used to remove fly ash and other fine particles from a gaseous stream. The particles are given an electric charge so that they are attracted to and are deposited on an oppositely charged electrode (usually a plate) where they are periodically removed.

ELEPHANT FIELDS Locations where sizeable oil and natural gas resources are located defined as having more than 100 million barrels of recoverable oil or one trillion cubic feet, or the equivalent in mixed resource. Since the Prudhoe Bay oil resource discovery, only a few have been found, mostly in the Overthrust Belt. Also see *Overthrust Belt.*

ELK KELP A genus of kelp (brown) found on the Pacific coast of the United States and Mexico which grows profusely, up to a length of 33 m (100 ft). Also see *Brown Algae; Kelp.*

ELKERITE A variety of bitumen formed through the slow oxidation of petroleum. Also see *Bitumen.*

EMBODIED ENERGY The total direct and indirect energy required for the production of goods and services. The indirect energy is that energy consumed to produce other than the direct energy items. It is not easy to determine the indirect energy because of the difficulty of determining the boundaries for the calculations. The embodied energy of steel is 62,770 kJ/kg.

EMERGENCY CORE COOLING A cooling system available for use in a nuclear reactor if the regular cooling system fails. The emergency core cooling system usually has more than one component each of which is designed to respond to a particular set of emergency circumstances. In water-cooled reactors, emergency core cooling is accomplished by injecting special water streams into the core.

EMERGENCY ELECTRIC POWER ADMINISTRATION (EEPA) An organization within DOE, formerly in USDI, which has authority over electric power in case of an attack on the United States or for other civil defense emergency.

EMF Electromotive force.

EMISSIVITY As applied to radiation heat transfer, the rate of emission of radiant energy from an opaque material due to the temperature, divided by the rate of emission by the same area of a black body at the same temperature. The total radiation from a black body is

$$q = \sigma A T^4$$

where

q = radiation, Btu/hr
σ = Stefan-Boltzmann constant, 0.173×10^{-8} Btu/ft^2 h °R^4

Equivalents:

0.173×10^{-8} Btu/ft^2 hr °R^4 = 5.71×10^{-5} ergs/cm^{2}sec K^4 = 56.7×10^{-12} kW/m^{3}K^4

A = area
T = absolute temperature

Emissivity is dependent upon the material, its surface characteristics, and temperature. Also see *Absorptivity.*

EMPIRICAL Refers to an equation, formula, curve, analysis, number, or relationship based on observation and experimental values. An empirical relationship is not necessarily supported by established theories.

ENDERGONIC A chemical reaction which occurs only with an input of energy, such as photosynthesis which requires an input of solar energy (or equivalent) to proceed. Also see *Endothermic; Exergenic.*

ENDOERGIC An energy absorbing process, the opposite of exoergic. Also see *Endothermic.*

ENDOTHERMIC A chemical reaction which requires an input of heat for the reaction to occur. Also known as endoergic. Also see *Endergonic, Exergonic; Exothermic.*

ENERGETICS The total energy relations and transformations of a system as in a chemical reaction, food system, ecosystem, etc.

ENERGY The capacity or ability to do work. Energy may be identified on the basis of type such as kinetic energy, potential energy, internal energy, and flow energy. Energy may also be classified according to source such as solar, gravity, geothermal, and nuclear, or as electrical energy, chemical energy, mechanical energy, biological energy, etc., (Figure E-1). Energy can neither be created nor destroyed, but can be converted to different forms.

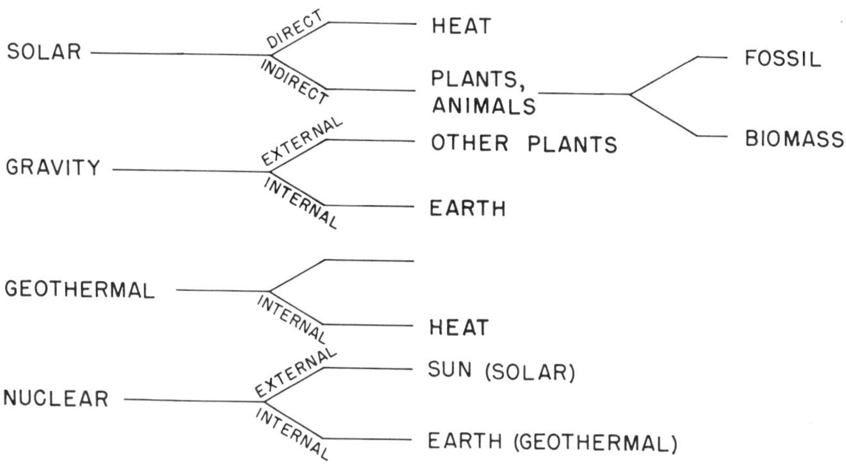

Fig. E-1 Energy.

ENERGY AUDIT An activity initiated by the DOE and state governments, in which information is obtained about a building, energy use is analyzed, and suggestions are made about reducing energy use. Size, use, history of energy use, hours of operation, vacation patterns, maintenance practices, etc., are evaluated. Modifications in facilities, their construction and use, are considered with the objective of decreasing the energy cost.

ENERGY CARRIER An agent by which energy is carried from one location to another. Although coal, gas, and oil are obvious or primary carriers, included also are heated air, water, steam, and possibly hydrogen and microwaves in the future.

ENERGY CONSERVATION Any deliberate action taken to reduce energy consumption such as avoiding use, finding alternative uses which require less energy, reducing dependence on the more costly less available sources, developing and utilizing more efficient energy using devices and processes.

The practice of effectively and efficiently using, rather than wasting, energy. Energy conservation can result from using less energy, using a source of lower energy availability, using waste heat, using a more efficient process, etc. The extent of conservation is greatly dependent on the economics— cost of energy, value of product, and manufacturing costs.

Examples of energy conservation are: Turning lights off when not in use; reducing level of illumination to lowest level needed; insulating homes, buildings, equipment, and pipes; utilizing waste heat; operating equipment and devices more efficiently, such as by lowering speeds, shortening time, loading to capacity as contrasted to partial loading, etc. Also see *Alternative Energy; Conversion; Efficiency.*

ENERGY CONSUMPTION A term used to represent energy utilized. Technically energy is not consumed, even though fuel is consumed, but is converted to a different form. Energy is neither created nor destroyed.

ENERGY CONVERSION The changing of energy from one form to another as, for example, the conversion of solar radiant energy into chemical energy in plants (photosynthesis), the conversion of chemical energy of fossil fuels to heat and products, such as during boiler operation or engine combustion, and the conversion of thermal energy to electricity. In all energy conversions some energy is lost during the conversion process. The percent converted is called conversion efficiency, Tables E-3 and E-4.

ENERGY CYCLE The name given by some authors to the carbon cycle. Also see *Carbon Cycle; Carbon Dioxide Cycle.*

ENERGY DENSITY The quantity of energy in a unit volume of a material, either in an energy source or fuel or power device. Usual units are J/m^3 and Btu/ft^3. Also see *Batteries; Heat Flux Density; Specific Density.*

ENERGY EFFICIENCY FOR ILLUMINATION The development of new methods of lighting has resulted in improved efficiency of converting electrical energy to light. Approximate efficiencies in lumens per watt of improved lighting devices are:

Carbon filament	3.5
Modern 100W tungsten	20
Tungsten-halogen	25
High-pressure mercury	50
Fluorescent	95
High-pressure sodium	100
Low-pressure sodium	180

TABLE E-3
Approximate Efficiency of Some Energy Conversion Processes

	Percent
Coal to electricity	
Chemical energy in coal, oil to steam	85-95
Thermal energy of steam to mechanical energy of a steam turbine	40-48
Mechanical energy to electricity	95-98
System of coal to electricity	32-42
Hydroelectric	
Mechanical energy in falling water recovery	85-90
Mechanical energy in turbine to electricity	95-98
System	80-88
Photovoltaic	
Solar energy to electricity by solar cells	15-25
Photosynthesis	
Solar energy in sunlight to chemical energy in plants	1-2

Also see *Efficiency*.

ENERGY EFFICIENCY RATIO (EER) In air conditioning the cooling effect, Btu/hr, divided by the energy used, watts, of an air conditioner as defined by the U. S. Department of Commerce. Usual values are between 4 and 12, with the higher number representing a more effective unit. A 'high efficiency' rating is 7.5 or higher.

ENERGY EQUIVALENTS The amount of energy in a given volume or weight of fuel or fuel source if all of the energy could be recovered as useful work. Thus, one pound of coal has the energy equivalent of 4 kWh, even though the one pound of coal would convert to only 1.15 kWh in practice (Figure E-2). Calorific values or heats of combustion,

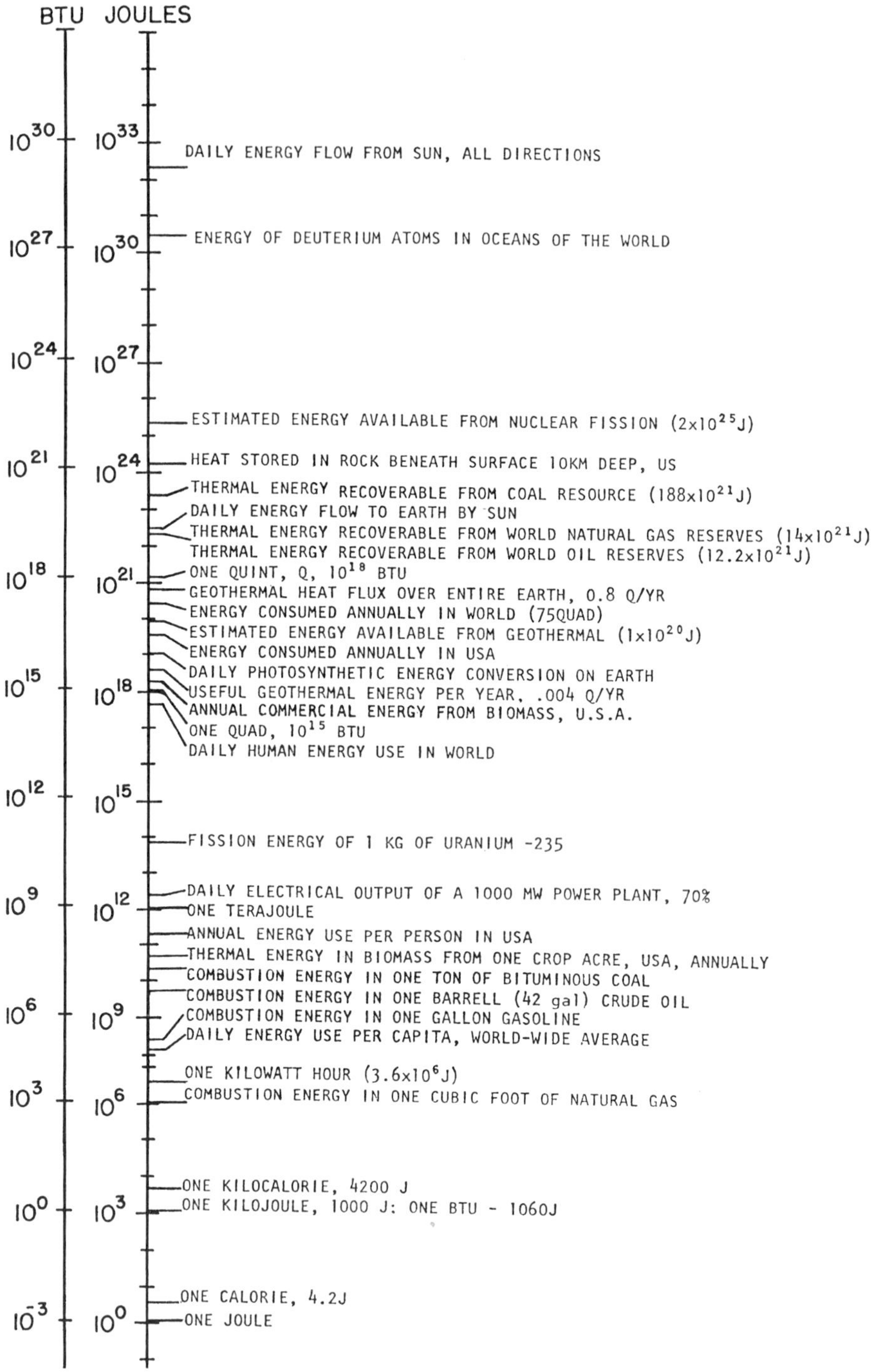

Fig. E-2 Energy Equivalents.

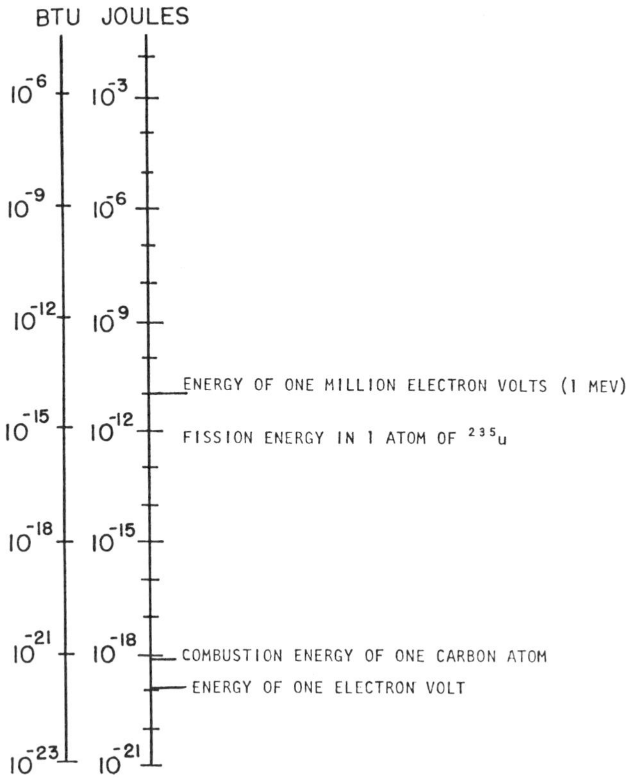

Fig. E-2 *(continued)*.

TABLE E-4
Conversion Efficiency of Various Fuels, Approximate Output/Input.

Initial material	Final product	Efficiency, %
Coal	Powdered coal	95
	Solvent refined coal (SRC), process	87
	Gas by gasifier	84
	Producer gas	75-80
	Synthetic oil	70 (42% energy efficient)
	Distillate	63
	Gasoline/diesel fuel	55
	Cogas plant efficiency	43-45
	Methanol	43
	Fischer-Tropsch liquids	37
	Methane	50-55 (max. 75%)
	Hydrogasification, process	65-70 (max. 90%)
	Electricity to users	28-30
Oil shale	Hydrogenated shale oil	75
	Distillate	67
	Gas, process	57-72
	Gasoline/diesel fuel	63
Biomass	Ethanol (grain)	62 (35% energy efficient)
	Ethanol (wood)	65
	Methanol (wood), process	30-37% thermal efficiency
	Hydrocarbon liquids	25
	Pyrolytic oil	35 (yield, 25%)
	Methane (fermentation)	33
	Low-/medium- Btu gas	70
	MSW gasification, process	66
	Gasification of wood, process	70
	Animal manure to gas, anaerobic	33
Petroleum	Clean crude oil	98
	Distillate	95
	Gasoline/diesel fuel	87-90
	Electric heating	17
	Electricity	30-38
	Methanol	65

From Hall, Carl W. (1981).

expressed in J/kg or J/m^3 or Btu/lb or Btu/ft^3, are energy equivalents. Also see *Calorific Values.*

ENERGY EXTENSION SERVICE A program of the DOE, established in 1977, charged with providing energy information to consumers and small businesses. The program is administered through state governments. A major objective of the Energy Extension Service is to encourage efficient use of energy. Also see *Energy Information Administration.*

ENERGY FOR CROP PRODUCTION In order to obtain production from the land various energy inputs are provided to enhance the use of solar energy. The major energy input in the United States for crop production is fertilizer, Table E-5, which is nearly twice the energy used in field operations.

TABLE E-5
Energy Used in Crop Production in the U.S.A.

Use	10^{12} kcal	10^{12} kJ
Fertilizer	122.5	512.6
Field operations	65.0	292.0
Irrigation	37.0	154.8
Drying	25.0	104.6
Pesticides	15.0	62.8
Machine manufacturing	7.5	31.4
Other	92.5	387.1

1 kcal = 4.1845 kJ = 3.968 Btu

Typical values of the energy for crop production inputs are:

For 1 kg equipment	87 x 10^3 kJ
1 kg nitrogen fertilizer	80 x 10^3 kJ
1 kg phosphorous fertilizer	14 x 10^3 kJ
1 kg potassium fertilizer	9 x 10^3 kJ
1 kg seed	30 x 10^3 kJ
1 kg 2,4-D	85 x 10^3 kJ
1 kg pesticide, fungicide	100 x 10^3 kJ
1 liter of fuel	40 x 10^3 kJ
1 kg irrigation water, Israel	200 x 10^3 kJ
1 kg irrigation, near water	.01 x 10^3 kJ

Also see *Energy Relationships; Energy Use in Food Production and Processing; Food System.*

ENERGY INFORMATION ADMINISTRATION (EIA) A group within DOE responsible for information and data on energy reserves, financial status of energy-producing companies, production, demand, and consumption. EIA is responsible for long-term analysis of energy trends. EIA is the clearinghouse for energy information and for publication and distribution of energy information. Also see *Department of Energy; Energy Extension Service.*

ENERGY LEVELS Discrete energy states in which nuclei, atoms, molecules, semiconductors, etc., can exist. Atomic spectra are produced by the light emissions accompanying electrons switching energy levels, or orbits, in atoms. Molecular spectra are similarly associated with the energy levels of electrons and ions moving in molecules. Nuclear magnetic spectra occur in transitions between nuclear spin energy levels.

ENERGY POTENTIAL OF CROP PRODUCTION SYSTEMS Crop production systems provide energy for food, feed, fiber, feedstock, fertilizer, and fuel. On the basis of metabolizable energy, sugar cane, potatoes, and cereal grains, in that order, have the greatest potential for producing food and fuel. Also see *C_4 Plants; Energy for Crop Production; Energy Relationships in Agriculture; Energy Use in Food Production and Processing; Food System.*

ENERGY RELATIONSHIPS IN AGRICULTURE Agriculture is a system which can produce more energy than it uses as commercial inputs because of the utilization of solar energy. The ratio of energy output/input for various products is:

Shifting crop	65
Rice and beans, peasant	41
Corn, manual	30
Cereal grains, sugar beets	2.5-5
Potatoes	1.6
Fruits and vegetables	0.1-1.0
Milk	0.4-0.5
Grass-fed beef	0.3
Eggs, intensive	0.2
Poultry, meat animals	0.1

Not to be overlooked is that food provides nutritous components other than energy. The relationship of farm output and energy in the United States is presented in Table E-6. Also see *Energy for Crop Production; Energy Potential of Crop Production Systems; Food Production and Processing.*

ENERGY SECURITY ACT An act of the Congress of the U.S.A. in 1980, called by many the Synfuels Act, which provides for a) Development of a synthetic fuels industry, including formation of a Synthetic Fuels Corporation, b) Utilization of vegetation to produce fuel, such as biomass for alcohol, with the objective of producing 10% of the estimated gasoline consumption by 1990, c) Filling the Strategic Petroleum Reserve (SPR) at a minimum rate of 100,000 bbl/day up to 500 million bbl in storage, d) Studying the formation of acid rain and effects of carbon dioxide from fuel combustion, and e) Other energy sources. Also see *Acid Rain; Synthetic Fuels Corporation.*

TABLE E-6
Energy Output and Energy Input for U.S. Farm Production Systems

	Output, 1959 = 100	Input, kJ x 10^{15}
1970	120	8.79
1965	112	7.53
1960	104	6.28
1959 (base)	100	6.07
1955	93	5.65
1950	85	4.81
1940	70	2.93
1930	62	1.67
1920	60	0.84

1 kcal = 4.1845 kJ = 3.968 Btu

ENERGY SOURCES During the 1975-1980 period, the U.S. annual consumption of commercial energy was 73-75 quads. The energy sources were approximately as follows:

Oil	45%
Natural gas	25%
Coal	20%
Nuclear	7%
Other	3%

ENERGY STORAGE See Heat Storage.

ENERGY USE IN FOOD PRODUCTION AND PROCESSING One estimate of the quantity of energy used in food production and processing in the U.S.A. is shown in Table E-7.

TABLE E-7
Energy Use in Food Production and Processing to the Consumer, U.S.

Sector	Energy use, kJ x 10^{15}	Percent
Production	0.24	3.8
Food processing	3.03	46.9
Transportation	0.18	2.8
Marketing and distribution	1.04	16.0
Household, etc.	1.97	30.5
	6.46	100.0

Note the low energy use in the production as compared to other sectors in the food system. Also see *Energy for Crop Production;*

Energy Potential of Crop Production Systems; Energy Relationships in Agriculture.

ENERGY USED IN DIFFERENT ECONOMIES The technological societies are greatly dependent on commercial energy, as shown in Table E-8.

TABLE E-8
Daily Consumption of Energy per Person, 1000 kJ

	Food	Industry, Agriculture	Commercial and Residential	Transportation	Total
Primitive man	8.4	-	-	-	8.4
Hunting man	12.5	-	8.4		20.9
Primitive agriculture	16.7	16.7	16.7		50.2
Advanced agriculture	25.1	29.3	50.2	4.2	108.7
Industrial	29.3	100.3	133.8	58.5	322.2
Technological	41.8	380.8	275.9	263.3	962.4

ENERGY USES DURING THE 1975-1978 PERIOD The annual consumption of commercial energy in the U.S.A. was 73 to 75 quads. The energy was used approximately as follows:

Industrial	30%
Transportation	25%
Household/commercial	20%
Electric power	25%

The world use of energy was 350 quads. In 1975 the world use of energy per person on a continuous basis was 2 kW, and in the U.S.A., 11 kW, for all purposes.

ENGINE A mechanical device for converting energy, usually thermal, into motion. If the primary energy is electrical, the device is called a motor. Engines are sometimes called motors. Engines may be classified as hot air, compressed air, external combustion (steam), internal combustion (gasoline), compression ignition (also internal) (diesel), rotating (turbine), reciprocating (pistons), rotary, etc.

ENHANCED RECOVERY OF OIL Following removal of crude oil from a reservoir, secondary recovery or removal of crude oil may be practiced by injecting gas and/or water and/or steam to increase recovery. Additional removal may be accomplished through tertiary recovery by flooding the reservoir with hydrocarbon solvents, polymers, high pressure carbon dioxide or nitrogen, or possibly by burning *in situ*.

ENRICHMENT The process by which the fraction of fissile isotope in a mixture of fissile and nonfissile isotopes is increased. Standard methods for enriching uranium include gaseous diffusion, gas centrifuging, and electromagnetic separation.

ENTHALPY Represented by H or h, the thermal energy of a substance with respect to an arbitrary reference level. The enthalpy is the sum of the internal energy, u or U, and the flow energy which is the product of the pressure, P or p, and the specific volume, v. Thus, $H = u + Pv$. For an ideal gas, $\Delta H = m\, c_p\, \Delta T$. Many energy calculations are concerned with the changes in enthalpy. The enthalpy tables are based on -17.8°C (0°F) for dry air, 0°C (32°F) for steam, and -40°C (-40°F) for refrigerants.

ENRICHED FUEL Fuel containing uranium that has a concentration of ^{235}U greater than the natural concentration (0.7%). It is necessary to enrich uranium in order to make a fuel that will sustain a chain reaction in a water-cooled core.

ENTROPY Represented by S or s, is a measure of the degree to which energy of a system is unavailable. The entropy of a body increases when its free energy decreases or when the disorder increases. When energy is lost by heat transfer, entropy increases. For thermal systems, entropy is defined as $\Delta S = \Delta Q/T$ for a reversible process, where ΔS is a small change in entropy, ΔQ is a small change in heat transferred to or from the system, and T is the absolute temperature at which ΔQ takes place. Also see *First Law; Free Energy; Second Law.*

ENVIRONMENT All of the physical, chemical, and biotic influences surrounding an organism and possibly affecting the life of the organism. An environmentalist is one who studies these relationships and/or one who is an advocate for maintaining or improving the environment. Also see *Ecology.*

EQUILIBRIUM The condition in which a balance exists between opposing forces or effects, such as moisture equilibrium, chemical equilibrium, temperature equilibrium, pressure equilibrium, etc.

EOC Enhanced oil recovery.

EOLIAN Caused by the wind. Also see *Aeolian.*

EON The largest unit of geological time which is 1,000 million years (Figure E-3). An era is shorter in time than the eon. Also see *Epoch; Era; Geologic Time; Period.*

EPA Environmental Protection Agency.

EPGA Emergency Petroleum and Gas Administration, USDI.

EPIGENE Those natural geological processes which occur near or

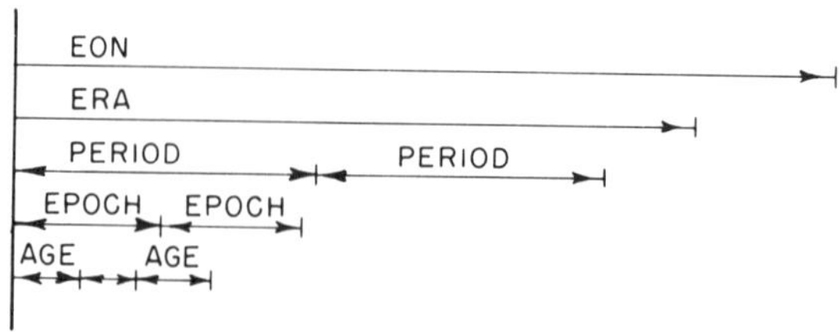

Fig. E-3 Geological time units.

at the surface of the earth, whereas hypogene refers to processes within the earth.

EPOCH A unit of geological time, with two or more epochs making a period. An epoch is made up of two or more ages. Also see *Eon; Era; Geologic Time; Period.*

EPRI Electric Power Research Institute.

EQUIVALENT DIRECT RADIATION (EDR) A relationship, not now generally used, between a heated surface and the rate of radiation from that surface as used for boilers, hot water heaters, and room heaters. The EDR for a steam radiator (215°F steam temperature) is 240 Btu/hr·ft^2 and for a hot water radiator is 180.

ERA A unit of geological time shorter than the eon; the era is subdivided into periods. The geologic eras are:

Cenozoic	65 millions years to present
Mesozoic	225 million to 65 million years ago
Paleozoic	570 million to 225 million years ago
Precambrian	Previous to 570 million years ago

Also see *Eon; Epoch; Geologic Time; Period.*

ERA Economic Regulatory Administration.

ERAB Energy Research Advisory Board (DOE).

ERCOT Electrical Reliability Council of Texas.

ERDA Energy Research and Development Administration (replaced by DOE).

EREP Energy related evaluation program.

ERG A unit of energy or work which results when a force of one dyne moves through 1 cm (0.01 m). 1 erg = 10^{-7} J. Also see *Joule; Work.*

EROS Earth Resources Observation Systems. Also see *Earth Resources Technology Satellite*.

ERR Energy-efficiency ratio.

ERRF Energy Resource Research Fund (Canada).

ERS Economic Research Service, USDA

ERTS Earth resources technology satellite.

ESCOE Engineering Societies Commission on Energy, Inc., Washington, DC.

ESP Electrostatic precipitator.

ESSERGY A general expression representing energy in a form essential for production of power or available to use for power. Also see *Exergy*.

ETHANE A colorless, odorless gas which is a hydrocarbon, C_2H_6. Ethane is one of the minor constituents found in most natural gas. It can be removed from natural gas and used as a feedstock for production of ethylene:

$$C_2H_6 + \text{steam } (34{,}400 \text{ kcal/kg}) \xrightarrow{\sim 900°C} C_2H_4 + H_2$$

The physical properties are: Boiling point, -88.6°C (-127.5°F); freezing point, -183°C (-297.9°F); critical pressure, 48.3 atm (90.3°F), flash point of -187.8°C (-306°F), and a specific gravity, 0.377. Also see *Ethylene*.

ETHANOL Also known as ethyl alcohol or grain alcohol, widely used as a beverage component, C_2H_5OH. Ethanol is now commercially manufactured from ethylene (C_2H_4), made from petroleum or from ethane, then ethylene from natural gas for non-beverage purposes.

$$C_2H_4 + H_2O \rightarrow C_2H_5OH$$

Ethanol for beverage is made by an aerobic fermentation of carbohydrates, principally grain, molasses, sugar, and fruit. Also see *Alcohol; Ethylene; Flash Point; Fuel Alcohol; Gasohol; Methanol; Net Energy Ratio*.

ETHLINE A mixture of ethanol and gasoline; ethline 10 represents 10% ethanol and 90% gasoline. Also see *Gasohol*.

ETHYLENE The name of a series of open chain hydrocarbons of which ethylene, C_2H_4, is the first in the series; a colorless gas with a sweet smell. Ethylene is one of the principal feedstocks in the chemical industry. The principal source is the cracking of petroleum, but it also is made from ethane in natural gas. Ethylene has a melting point of -169°C (-272°F); boiling point of -103.7°C (-154.7°F). Also see *Ethane*.

ETSI Energy Transportation Systems, Inc. (Pipeline.)

EUTROPHIC The deterioration caused by an overloading of a water body (stream, river, lake, etc.) by excessive growth of organic matter. Eutrophication results from the addition of rich nutrients from fertilizer or other nutrient run-off, factory, home, or city effluents. With excessive nutrients the water body loses its ability to maintain life support systems primarily due to the depletion of oxygen.

EV Electric vehicle.

eV Electron volt.

EVAPORATION The change of a liquid to a gas which occurs at the surface of a liquid. The evaporation occurs as excited molecules at the surface of a liquid escape. The rate of evaporation for a particular liquid depends on its temperature and the vapor pressure of the gaseous phase in contact with it. The latent heat of vaporization is the quantity of heat per unit mass required to change a material from the liquid to the gas phase at a particular temperature. The latent heat of vaporization of water at sea level is 2257 kJ/kg (970.4 Btu/lb) at 100°C (212°F). Boiling may be considered as evaporation throughout the liquid. Evaporation is a synonym for vaporization. Also see *Latent Heat; Vaporization.*

EVAPOTRANSPIRATION The process by which water is lost from the land, including plant surfaces, by evaporation and by transpiration from plants. Evapotranspiration rates are greater in areas of high solar energy intensities.

EXCURSION An unplanned transient operation of a nuclear reactor. The term "excursion" sometimes is used for a planned transient, but the usual implication is an accident situation.

EXERGONIC A chemical reaction which gives off energy, such as in the respiration process, the opposite of endergonic. Also see *Endergonic; Endothermic; Exothermic.*

EXERGY Energy in a substance in relation to the surroundings (for water lifted above body or for air temperature above surroundings) in contrast to using the lowest energy level (such as sea level or absolute zero temperature for thermal application).

Exergy is the energy available for useful work under specified conditions. The useful energy, or exergy

$$E_{ex} = \frac{T_1 - T_2}{T_1} E_{en}$$

where
T_1 is the temperature of the substance and T_2 is the temperature

of the surroundings in which the Carnot efficiency is $\frac{T_1 - T_2}{T_2}$. Also see *Anergy*.

EXOERGIC A process which releases energy, or exothermic, the opposite of endoergic.

EXOSPHERE The layer of the atmosphere outside the ionosphere which has a low density. The exosphere is defined as that part of the atmosphere where a molecule moving outward has a 50% chance of escaping. Also see *Atmosphere; Ionosphere*.

EXOTHERMIC A term used to identify chemical reactions in which energy is liberated during the reaction. The burning of carbon as in coal produces carbon dioxide and heat. The burning of one kilogram of carbon gives off 33,890 kJ and is an exothermic reaction. Also see *Combustion; Endergonic; Endothermic; Exergonic; Exoergic*.

EXPLOITATION WELL A well drilled to determine the potential oil and/or gas field and the characteristics of the fluid.

EXPLOSION, GASES A violent reaction which is the result of rapid combustion in which the heat cannot be carried away as rapidly as it is produced. Air (oxygen) - fuel (gas) ratios determine to a large extent the possibility of explosion, assuming that ignition and mixing are favorable (Table E-9). Safety to people and equipment is a primary concern in design and operation of energy systems so as to minimize the potential for explosion.

TABLE E-9
Limits of Flammability in Air, Percent

Gas	Lower	Upper
Acetylene	2.5	80
Blast furnace gas	35	73.5
Butane	1.9	8.5
Carbon monoxide	12.5	74.2
Coal gas	4.8	33.5
Ethane	3.0	12.5
Hydrogen	4	74.2
Methane	5.3	14
Natural gas	4.8	13.5
Producer gas	20.2	71.8
Propane	2.2	9.5
Water gas	6.9	70.5

Also see *Limit of Flammability*.

EXPLOSION, DUST A violent explosion which is the result of rapid combustion of dust and particles in the atmosphere in which heat cannot be carried away as rapidly as it is produced. For dusts the limits on explosion characteristics are described by ignition temperature of the dust cloud, minimum and maximum explosive concentrations. Of these the maximum concentration for explosion is not well defined. Some representative values for different materials are listed in Table E-10.

TABLE E-10
Explosive Concentration of Dust

Solid material	Ignition temperature °C	°F	Explosive concentration minimum, kg/m^3 or oz/ft^3
Cellulose fibers	450	806	0.035
Coal, 40% volatile	610	1130	0.050
Lignite	450	707	0.03
Peanut hulls	460	860	0.045
Peat, sphagnum	460	860	0.045
Walnut shells	450	707	0.03
Wheat straw	470	878	0.055
Wood	430	806	0.035

Based on U.S.B.M. RI 5753 and RI 5971.

Also see *Explosion, Gases.*

EXPOSURE A measure of the ionization produced in air by X- or gamma radiation. It is the sum of the electrical charges of all of the ions of one sign produced in air when all electrons liberated by photons in a volume element of air are completely stopped in the air, divided by the mass of air in the volume element.

EXTERNAL COMBUSTION A heat engine in which the combustion chamber is outside the engine, such as a steam engine. With the combustion outside of the engine improved combustion characteristics are possible as a result of continuous combustion. The steam turbine and Stirling engine are other examples of external combustion engines. Also see *Steam Turbine; Stirling Engine.*

EXTRA-HIGH VOLTAGE (EHV) Electric transmission systems below 1,000 kV, but above the secondary voltages, usually 500,000 or 750,000 volts.

EXTRACT HYDROGENATION The process of hydrogenation of an extract of coal to produce synthetic crude oil. The process is the basis of the CSF process to utilize high-sulfur coal to produce acceptable low sulfur products. Also see *Consol Synthetic Fuel; Hydrogenation.*

EXTRACTION-HYDROGENATION The extraction of sulfur carried out in the presence of hydrogen either as a gas or derived by the transfer of hydrogen donor solvent. Also see *Exxon Donor Solvent; Hydrogenation.*

EXXON DONOR SOLVENT (EDS) PROCESS A process for removing impurities from coal to produce a clean soluble coal, char, and gases, which uses a single plug-flow reactor and external catalytic hydrogenation to maintain the solvent (Figure E-4). Also see *H-Coal; Solvent Refined Coal.*

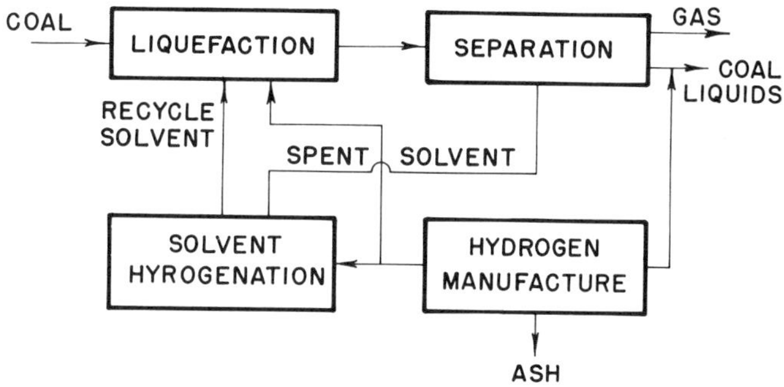

Fig. E-4 Exxon donor solvent process *(AIChE, 1980).*

F

FACECORD OR FACE CORD   A pile of logs or wood eight feet long and four feet high, but with logs shorter than four feet required to make a cord. Also known as a short cord, pick-up cord, stove cord, fitted cord, or fireplace cord. Also see *Cord*.

FAHRENHEIT TEMPERATURE SCALE   A uniformly graduated scale in the English system of units for which 0°F was established as the melting point of ice in a sal ammonica and water solution, and 96°F was mistakenly used as human body temperature. Based on the scale established in this way it turns out that 32°F is the freezing point of water, and 212°F the boiling point. To convert to °C, subtract 32 from °F, then multiply times 5/9. Also see *Celsius*.

FAIL SAFE   Phrase describing a process or system which will always cease functioning in a safe condition in case of a power failure or other malfunction. Generally the process or system in normal operation can be in any of two or more states, some of which would not be safe as terminal conditions. The fail safe system or process is designed not to permit the non-safe states to be terminal states.

FAIL-SAFE CONTROL   A control designed so that a malfunction will stop the operation of the device or equipment in a safe position. Examples are a burner, vacuum pump, nuclear control, or pressure pump.

FALLOUT   Generally refers to the fallout of radioactive materials from the atmosphere, such as from nuclear explosions; can also refer more generally to the fallout of other materials such as solids from power plants, industrial operations, volcanoes, etc. For radioactive fallout, fission products are most significant, particularly strontium-90, which accumulates in calcium in the body and has a half-life of 25 years. Also see *Daughter Fission; Half-life; Nuclear Radiation*.

FAST BREEDER REACTOR (FBR)   A nuclear reactor that produces more fuel than it consumes and that derives most of its energy from nuclear fissions initiated by neutrons that have not slowed down to the thermal range. The light water reactors built in this country are not fast reactors but thermal reactors, i.e., the energy is produced by neutrons that have slowed down to the thermal range. Also see *Nuclear Fuel Cycle*.

FAST FLUX TEST FACILITY (FFTF) A test reactor built on the Hanford Reservation in the state of Washington. This reactor is the principal facility for engineering tests of the components of commercial liquid metal-cooled fast breeder reactors.

FAST NEUTRONS Neutrons that have kinetic energies greater than some lower limit, usually taken to be a few hundred kiloelectron volts.

FATHOM A unit of length measurement used for depth, particularly for marine applications, equal to 1.829 m (6 ft).

FBC Fluidized bed combustion.

FBH Fluidized bed hydrogenation.

FBR Fast breeder reactor.

FCG Fuel cell generation.

FCR Full core reserve.

FDA Food and Drug Administration.

FEA Federal Energy Administration.

FECAL ENERGY The gross energy of the feces of people and animals which consists of the energy content of the undigested food and the metabolic portion of the feces. Also see *Apparent Digestible Energy; Metabolizable Energy*.

FEDERAL ENERGY GUIDELINES A compilation of energy regulations of DOE, published weekly in loose-leaf form by the Commerce Clearing House, a commercial publisher. Also see *Code of Federal Regulations; Federal Register*.

FEDERAL ENERGY REGULATORY COMMISSION (FERC) A five person group appointed by the president with the consent of the Senate members, serving four year terms. FERC establishes interstate rates of gas and electricity and issues permits for non-federal hydroelectric facilities, including dams. It was established in 1974 as the successor to the Federal Power Commission.

FEDERAL POWER COMMISSION (FPC) A federal commission no longer in existence, which was established in 1920 to regulate hydroelectric development and interstate trade of electricity and natural gas. Many of the functions of the former Federal Power Commission have been assumed by the Federal Energy Regulatory Commission (FERC).

FEDERAL REGISTER A publication of the United States government in which all government regulations are published as a part of rule-making procedures. The Federal Register is published every working day of the federal government. It is also used to announce meetings, to present significant and non-significant regulations, decisions on cases, availability of reports, and preliminary copies of major regulations. The Federal Register is published by the Government Printing Office, and is available from the National Archives and Records Service, 8th and Pennsylvania Avenue, N.W., Washington, DC 20408. Also see *Code of Federal Regulations; Federal Energy Guidelines.*

FEEDSTOCK The raw material supplied to a chemical process. Fossil fuels, for example, are feedstocks for gaseous fuels, chemicals, etc.

FEED MATERIALS (nuclear) The fresh fuel material that is introduced into a nuclear reactor to replace used fuel that is removed.

FEO Federal Energy Office.

FERC Federal Energy Regulatory Commission.

FERMENTATION The process of anaerobic respiration resulting in the breakdown of organic materials with the production of carbon dioxide and other products such as alcohol. These chemical changes occur by the action of yeast or bacteria in materials containing starch or sugar.

FERTILE MATERIAL Material that can be converted into nuclear fuel by some nuclear reaction such as neutron absorption.

FFB Fuels from biomass.

FFTF Fast Flux Test Facility.

FGD Flue gas desulfurization.

FIERY MINE A coal mine from which methane is produced in significant quantities. Appropriate ventilation is required to maintain working conditions. The gas may be recovered for fuel.

FILM BADGE A badge, containing a film sensitive to nuclear radiation, that is worn by personnel who may be exposed to radiation. The most common badges are sensitive to beta and gamma radiation, which are separately determined using different film segments. Badges that measure neutron exposure also are available.

FILTER Device used to remove components, usually undesirable, from gases and liquids. The size of the pores in a filter determines the size of particles retained. The air supply to a controller, the fuel to a burner, and air for ventilation are examples of gas streams that are often filtered. Absolute filters will remove 99.9% of particles, including microorganisms, that have diameters of 0.3 micron or larger. Filters are classified as roughing, medium efficiency, high efficiency, and ultra high or absolute. In general, the greater filtration requires a larger pressure drop across the device and is more expensive to operate. Coarser filters may be called strainers. Also see *Baghouse*.

FIME Fissions per initial metal atom.

FIRE CLASSIFICATION (NFPA) The National Fire Protective Association identifies three letter classifications of fire, based on their origin: A, solid materials which must be heated to ignition point; B, flammable materials, usually liquid and gas; and C, those in electric equipment.

FIRE POINT The temperature of an oil at which enough vapor is released to burn continuously. Also see *Flammable; Flash Point; Ignition Point*.

FIREDAMP The name given to gas, principally methane, which may collect and form an explosive mixture with air. Also see *Explosion, Gas; Marsh Gas; Methane*.

FIRST LAW OF THERMODYNAMICS The physical law which states that energy is conserved during a process. Although heat may be transferred from one place to another during a process there is no net gain or loss in total energy. For a system which undergoes a cycle, the net heat transfer across the system boundary after each complete cycle is equivalent to the net work done by the system. Also see *Second Law of Thermodynamics*.

FISCHER-TROPSCH PROCESS A process of hydroliquefaction of coal in which liquefaction is carried out following gasification, and in which coal is converted to gas, liquids, and char, with about 40 to 50% efficiency. The first step of the present Fischer-Tropsch process is a Lurgi gasifier which produces a synthesis gas from coal in a steam/oxygen mixture in a high pressure environment.

FISSILE Capable of fissioning. The common fissile nuclei are ^{235}U and ^{239}Pu, but many other nuclei possess this property. Some nuclei fission spontaneously, but more often neutron absorption initiates the fission.

FISSION A nuclear reaction during which a nucleus divides into two or (rarely) three principal parts, called fission products, plus possibly other lighter fragments, especially neutrons. Nuclides capable of undergoing fission are called fissionable or fissile. Some nuclei fission spontaneously, but most fissile nuclei must absorb a neutron before fissioning. A fissile nuclide is capable of fissioning in a wide variety of ways so that many pairs of fission products are possible.

In the fission process mass is converted to energy. The fission of ^{235}U yields 2.5 million times as much energy as burning the same mass of carbon. Also see *Chain Reaction; Fusion; Nuclear Reactor.*

FISSION PRODUCTS The main fragments produced by the fission (dividing) of a nucleus. Each fission usually produces two fission product nuclei. There are many possible pairs of fission products that can result from fissions of a given type of initial nucleus. Also see *Fission.*

FISSIONABLE MATERIAL Material whose nuclei are capable of fissioning.

FISSIONS PER INITIAL METAL ATOM (FIME) A measure of the amount of burnup achieved in the fuel of a nuclear reactor. As the name indicates, the measure used is the number of fissions produced in the fuel per metal atom, including ^{238}U as well as ^{235}U, initially present in the fuel.

FISSIUM Fissile material artificially mixed with fission product elements to simulate irradiated fuel.

FIXED BED A chamber in which the product(s) for combustion, drying, or processing is held in a stationary position during treatment. In contrast, in a fluidized bed the product moves during treatment, providing a more rapid response.

FIXED CARBON The carbon in a material, especially coal, which is not removed with the volatile materials during heating, usually expressed as a percentage. Also see *Free Carbon.*

FIXED OIL See Oil.

FIXED POSITION A neutron absorbing material located in a fixed position in a nuclear reactor. The alternative to a fixed position is a movable position, such as the neutron absorber in a control rod, that can be moved into or out of a reactor.

FLAME The visible effect from radiation during combustion. The flame results from burning of the soot particles and the gaseous combustion products. The temperature of a flame varies as shown in Figure F-1. The temperature and color of flames vary with the gas or other element involved (Table F-1). A sodium flame in oxygen is yellow; copper, green; and phosphates, blue.

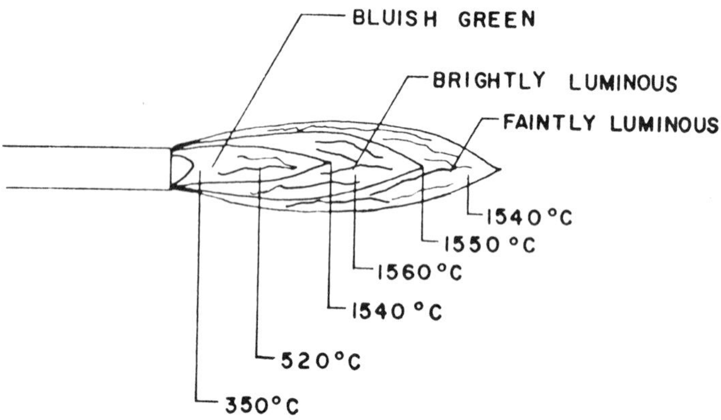

Fig. F-1 Flame-color and temperature distribution.

FLAMMABLE Able to support combustion, describing, for example, a fuel. Because of the possibility of misunderstanding the meaning of the word inflammable, a term previously used for this purpose, flammable is now used to a greater extent commercially. Also see *Explosion, Gases; Inflammable; Limits of Inflammability.*

TABLE F-1
Flame Temperature of Various Gases, Approximate

	°C	°F
Butane, C_4H_{10}	1875	3440
Candle	1500	2732
Carbon monoxide, CO	2400	3800
Coal gas and air	2000	3632
Cyanogen	2330	4226
Hydrogen, H_2	2350	3700
Methane, CH_4	1850	3400
Natural gas	1900-2300	3700
Oxyacetylene	3340	6000
Propane, C_3H_8	1950	3500

FLARING The commercial or industrial practice of venting and burning the unwanted gas associated with oil wells. The gas produced along with the oil was once wasted by flaring but now regulations are in effect prohibiting flaring in most cases, without special permit. The produced gas, which is mostly methane, can be used as a gas or can be liquefied and used commercially.

FLASH Rapid generation of a heated fluid changing from a liquid to a gas.

FLASH POINT The lowest temperature at which a liquid produces enough vapor to burn in air when ignited at a given pressure (Table F-2). Important for storage, transportation, and safety considerations for fuels. Flash point also refers to the temperature at which lubricants for compression refrigeration units, such as used in freeze dryers, vaporize. Also see *Fire Point; Ignition*.

FLAT PLATE COLLECTOR A flat solar panel, which may or may not be covered with glass and which absorbs heat from the sun (Figure F-2). When covered with one or more layers of glass, the incoming short waves from the sun pass through the glass and are absorbed and changed to long waves in the form of heat which is trapped between the glass and the collector. The heat is used to increase the temperature of air or a liquid, usually water, flowing through space or pipes in the collector. The flat plate collectors are usually, but not always, mounted in a fixed position facing the sun during most of the day. Flat plate collectors are particularly applicable and useful for heating water. Also see *Focusing Collector; Solar Energy; Solar Energy Collector*.

TABLE F-2
Flash Temperatures of Liquid Fuels and Oil

Benzene, C_6H_6	-11°C (12°F)
Ethanol, C_2H_5OH	12°C (54°F)
Fuel Oil	
No. 1	38-74°C (100-165°F)
No. 2	43-88°C (110-190°F)
No. 3	43-110°C (110-230°F)
No. 5	55°C (130°F)
No. 6	65°C (150°F)
Lubricating oils	163-218°C (325-425°F)

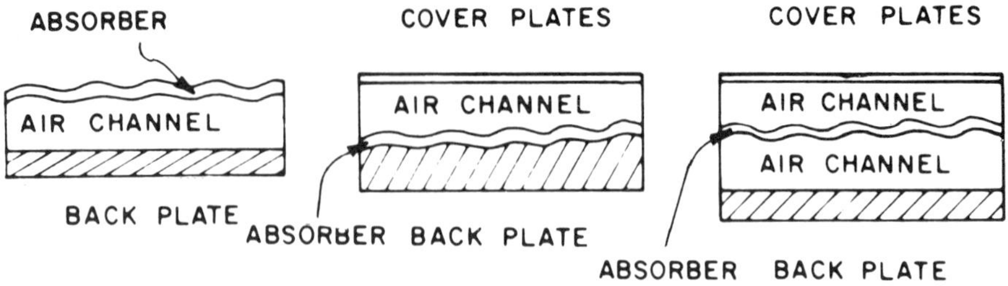

Fig. F-2 Flat plate solar energy collectors

FLOAT COAL A collection or assemblage of fine pieces of coal produced by natural forces such as wind and rain.

FLOCCULATION The combining of small particles into layer clumps, as with flocculation of coal to produce larger clumps. Flocculation can be accomplished by chemical treatment. Alum, ferric iron, and organic polymers are used as flocculents.

FLUE GAS A mixture of gases emerging from a stack, chimney, or flue and resulting from combustion. Flue gases consist of oxides, carbon monoxide, carbon dioxide, water vapor, sulfur dioxide, particulates, etc. An analysis of flue gas, as with an Orsat apparatus, provides information for evaluating efficiency of combustion. Also see *Combustion; Pollution*.

FLUID CATALYTIC CRACKING; FLUID-BED CATALYTIC CRACKING The process by which crude oil or heavy oil is converted to gasoline and distillate by thermal heating in the presence of a catalyst. One category of catalyst is the hydrated alumina silicates. Catalytic cracking provides 40% of C_4 fraction as compared to thermal cracking which produces only 17% of C_4 fractions.

FLUID HYDROCRACKING The process for breaking down heavy distillates and residual oils at pressures of 800 to 2500 psig in the presence of hydrogen. The hydrocracking processes handle a larger range of materials than catalytic cracking, and produce gasoline products of lower octane number. Also see *Hydrocracking*.

FLUID HYDRODESULFURIZATION A fluidized bed process for removing sulfur in the presence of hydrogen and a catalyst such as cobalt or molybdenum on a porous alumina base. Also see *Desulfurization*.

FLUIDIZED BED; FLUID BED A chamber in which the products being processed are held in suspension or moved above the bed by the air or gas stream moving into the chamber. In the fluidized bed the product moves during treatment providing mixing and rapid response.

Fluidized bed devices are used for: Boilers, catalytic cracking, coking, hydrocracking, combustors, dryers, gasifiers, methanators, and reactors.

FLUX The rate of flow per unit cross section, such as J/m^2s (Btu/ft^2hr) for heat transfer; lines of force per unit area for electric and magnetic fields and power incident per unit area for radio waves.

FLY ASH The small noncumbustible particles or particulates which leave the combustion area of a coal-fired boiler and which could be a major cause of air pollution. There is considerable variation in the amount and kind of fly ash produced depending on coal and design and operation of the combustion unit. As much as 10% of the coal may become unburnable fly ash. Centrifugal separators, gravity collectors, and electrostatic precipitators are used to remove particulates. Standards have been established by the 1970 Clean Air Act in the United States. The average concentration over a 24 hour period should be no more than 260 micrograms per cubic meter of air.

FLYWHEEL STORAGE A method of storing kinetic energy in a rotating flywheel. Energy from braking a moving vehicle can be put into rotating a flywheel, from which energy can be taken to propel the vehicle.

FOCUSING COLLECTOR A device for collecting solar energy and concentrating it into a smaller area. The focusing collector has the cross section of a parabola, either as a dish collector or line-focus. Mirrors may be used to focus energy on the absorbing surface. The collector can be mounted on the focal line and turned facing the sun to get maximum energy collection and temperature increase. Also called concentrating collector. Also see *Concentrating Ratio; Flat Plate Collector; Solar Energy.*

FOOD AND ENERGY COUNCIL 909 University Avenue, Columbia, MO 65201.

FOOD CYCLE As applied to the ocean, the production, consumption, and decay of food, and the energy relationships involved. Bacteria transform the products of decay into nutrients which are useful for plants which are available for marine animals. Also see *Energy for Food Systems; Food Systems.*

FOOD SYSTEM The three major food systems are traditional or primitive, which involves primarily production; transitional, which involves production and processing; and mechanical or modern, which involves production, processing, and preparation. To move from the traditional to the modern mechanical system, external or commercial energy must be added.

In primitive or traditional food production, people on the land produce and utilize their own food and fiber with little of their production leaving the land. As modernization occurs, people move from the farm to industry. The farm is left with fewer hands so that farming has to become more productive to meet not only the food and fiber needs of those living on the farm but also those working in industry in the cities. Food and fiber flow from the farm to the cities and industry and the farm became more dependent on goods and services from industry. A principal external need is energy, either directly as a fuel or indirectly in the form of nutrients, environmental control, and water supply. Countries with rapid industrial development and low population density have undergone extensive mechanization with high use of external energy.

Highly mechanized countries can usually be identified as those with high horsepower per hectare. However, other factors, such as mechanical aids for field production, water control, materials handling, storing and processing become important for producing more and saving more of the product.

The energy efficiency of subsistence crop production is high but the output per worker per hour is 11 to 19 kilojoules per worker. By contrast the output for cereals in an industrial economy is 2800 kilojoules per worker, or 200 times as great in industrial agriculture as for subsistence crop production.

Using external energy balances is not a proper method of measuring the quality of life, productivity, etc., in rural subsistence economics, often considered poor. In rural subsistence agriculture the people produce their own inputs and consume their own outputs. Materials are recycled within the unit, with the influx of outside solar energy as the primary outside or external energy source. Under such conditions a recycling ratio of 62 to 73% of materials occurs on 1.5 to 3 hectare subsistence units. *(Axinn and Axinn.)* Also see *Energy Use in Food Production and Processing; Energy Used in Different Economies*.

FOOT-POUND A unit of energy or work equal to the work done in lifting one pound through a vertical distance of one foot. Also see *Energy; Erg; Work*.

FORCE The physical agent which brings about a change in the momentum of a body and is directly proportional to the product of the mass, m, and acceleration, a, of a body, $F = ma$, where $m = w/g$ (Table F-3).

FOSSIL The replacement of an organism or part of an organism covered by land/water, by mineral matter, such as with silica, based

TABLE F-3
Force Units

System	Force	Mass	Acceleration
cgs	1 dyne	1 gram	1 cm/sec^2
mks	1 newton	1 kg	1 m/sec^2
fps	1 poundal	1 slug	1 ft/sec^2
SI	1 newton (N)	1 kg	1 m/s^2

$F = m \times a$

on time, temperature and pressure of the surroundings. Also see *Fossil Fuels.*

FOSSIL FUELS Coal, lignite, oil, natural gas, tar sands, and oil shale; used primarily for burning, consisting primarily of carbon and hydrogen. Fossil fuels developed from plants and animals which became embodied in the earth and changed to their present forms over an extended period of time.

FOULING See Biofouling.

FPC Federal Power Commission (replaced by FERC).

FPR Federal procurement regulations.

FRACTIONAL DISTILLATION The separation and collection of components or fractions of materials which vaporize upon heating. The factions in heating of crude oil from higher to lower boiling points are light gasoline and petroleum gas, naphtha, kerosine (200°C), light gas oil, heavy gas oil (300°C), and residual oil.

FRACTIONS The portions of a petroleum product produced by distillation or cracking with the more volatile fractions removed first. Naphtha and various gas-oils are examples of the lighter fractions. Also see *Fractional Distillation.*

FRACTURING The process of cracking and fissuring a gas reservoir to increase the production. Also see *Proppant or Propping Agents.*

FRANCIS TURBINE A reaction hydraulic turbine applicable for heads of 30 to 300 m (100 to 1000 ft). Also see *Hydraulic Turbine; Kaplan, Pelton.*

FRASCH PROCESS A method of mining sulfur using superheated hot water to melt sulfur which is then pumped out of the well. Also see *Claus Process.*

FREE CARBON The carbon, usually expressed in percentage, removed during heating in contrast to the fixed carbon which is not removed. Also see *Coal Classification; Fixed Carbon*.

FREE ENERGY The difference between the enthalpy and the product of entropy and absolute temperature of a thermodynamic system. Also called Gibb's free energy and thermodynamic potential.

$$\text{Free energy, } G = H - TS$$
$$= U - TS + pV$$

where

H = enthalpy, H = U + pV
U = internal energy
T = absolute temperature
S = entropy
p = pressure
V = volume

Sometimes used to designate the Helmholtz free energy, A = U - TS. Also see *Enthalpy; Entropy*.

FREE MOISTURE OR FREE WATER The water in a substance not held by hygroscopic forces. The water can be removed by mechanical methods or gravity, such as draining, centrifugal force, and pressing. Some authors refer to free moisture as that moisture in a product above the equilibrium moisture content, $M - M_e$. Some authors define free moisture content as including unbound water. Also see *Moisture Content Representation*.

FREEZING POINT The temperature at which a liquid or gas changes to a solid without a change in temperature, releasing heat, known as the heat of fusion, at a given pressure. Representative values are:

	Water	Nitrogen	Oxygen
Temperature, °C	0	-210	-219
°F	32	-346	-362
Latent heat			
kJ/kg	334.9	25.5	13.7
cal/g	79.7	6.1	3.3
Btu/lb	144	10.9	5.9

Also see *Latent Heat*.

FREIGHT EFFICIENCY A measure of the weight-distance (ton-miles) which freight can be moved for a given quantity of fuel. Pipelines have the highest efficiency for moving liquid materials. Approximate freight efficiencies are:

	ton-miles/gal	tonne-kilometers/liters
Oil pipelines	295	115
Waterways	250	97
Railroads	190	74
Trucks	60	23
Aircraft, heavier than air	10	4

Also see *Passenger Efficiency*.

FRICTION The resistance to movement of two objects in sliding contact with each other. For solid bodies, the coefficient of friction between two surfaces is the ratio of the parallel (to the surface) force required to move one over another to the normal force of one body against the other. The coefficient of friction relates to calculations involving handling and storing of liquids and granular materials (Table F-4). The coefficient of friction is the tangent of the angle of friction, ϕ. One body will just slide over another at the angle of friction, ϕ. Values of the coefficient of friction will change depending on temperature, moisture content, and surface conditions. The coefficient of friction for initiating movement, known as the static coefficient, is larger than the sliding or kinetic coefficient of friction. The maximum angle from the horizontal which a bulk material will assume when piled is the angle of repose or internal angle of friction. The approximate angle of repose for several materials is:

Ashes	40°	Coke	30-45	Sand, dry	25-35
Cement	40°	Limestone, crushed	37		
Coal, hard	27°	Liquid	0		
Coal, soft	35				

TABLE F-4
Friction, Coefficients of (Estimates only)

	Steel	Concrete	Wood (smooth)
Ashes	0.60		
Coal, hard	0.28	0.50	0.47
Coal, soft	0.32	0.70	0.70
Corn	0.41	0.44	0.36
Rice	0.41	0.52	0.44
Sand, dry	0.32	0.57	0.57
Sawdust, wet	0.60		
Steel, dry	0.15-30	0.30-0.7	0.20-0.62
Wheat	0.37	0.42	0.31-0.425
Wood	0.50-0.62	0.4	0.20-0.54

Also see *Angle of Repose*.

FRICTION HEAD The loss or drop in head of flowing fluid through a hydraulic device, in length units, cm, m (in, ft) or conventional pressure units, kPa (psi).

FUEL A substance which will release energy as a result of chemical change, e.g., combustion, or radioactive release, Figure F-3, as a source of power and work. Also see *Energy; Various Fuels.*

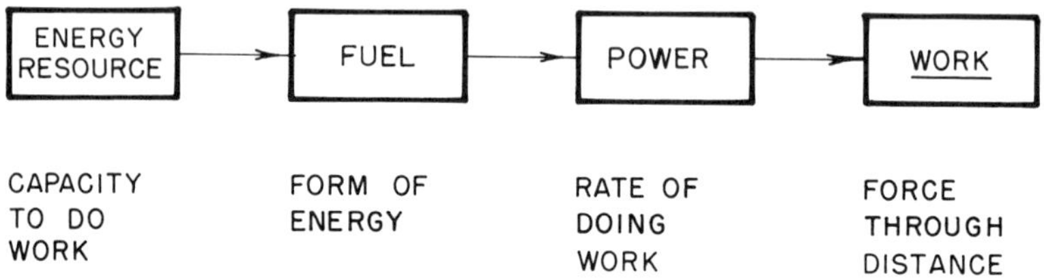

Fig. F-3 Fuel in energy used to do work.

FUEL ALCOHOL Alcohol used for fuel. Also see *Alcohol; Anhydrous Alcohol; Hydrous Alcohol.*

FUEL ASSEMBLY (nuclear) A group of fuel elements assembled and held in a frame so that they can be handled together as a unit. The use of fuel assemblies simplifies the handling of fuel in a nuclear reactor.

FUEL BURNOUT The loss of fuel in a nuclear reactor that takes place as the fissile nuclei fission.

FUEL CELL A device in which chemical energy is converted directly to electrical energy by supplying fuel, such as hydrogen or methane, to electrodes which are in an electrolyte. As a result current flows establishing an electric source. A hydrogen-oxygen reaction in a fuel cell is the reverse of the electrolysis of water. In an H-O fuel cell, hydrogen and oxygen combine to produce water and an electric current. Also see *Bug Battery.*

FUEL CHANNEL (nuclear) A channel in the core of a nuclear reactor into which a fuel assembly can be inserted. The channel is usually bounded by holes in structural templates in the core.

FUEL CLUSTER (nuclear) A fuel assembly.

FUEL ELEMENT The unit fuel module in a nuclear reactor. The typical fuel element consists of a fuel-bearing region surrounded by a thin layer of metal cladding that isolates the fuel from the reactor coolant.

FUEL GAS A general name given to a gas which can be distributed by or through a pipeline for fuel purposes. Fuel gases include: Natural gas, methane, propane, refinery gas, coal (low- and medium-Btu) gas. Fuel gas is primarily gaseous carbon and hydrogen compounds but may also include carbon monoxide, hydrogen, etc. Fuel gases generally have a lower ratio of carbon to hydrogen atoms than liquid and solid hydrocarbon fuels. Also see *Biogas; Coal Gas; Methane; Natural Gas; Producers Gas; Propane.*

FUEL OILS The heavy fractions of relative low boiling point which result from normal petroleum refining processes. Fuel oils have a higher boiling range than kerosene. Fuel oils are generally distillates or residuals. There are many grades of fuel oil. The heavier fuel oil must be preheated for combustion and is used in commercial applications such as boilers and furnaces. Also see *Flash Point.*

FUEL PELLET (nuclear) A small pellet made from a fuel-bearing material. The common fuel element in a light water reactor consists of a column of uranium oxide fuel pellets filling the central space inside a hollow cylindrical cladding.

FUEL REPROCESSING (nuclear) A process that extracts the unused fuel from spent fuel elements so that it can be re-used.

FUEL ROD The assembly in a nuclear reactor consisting of the fuel (uranium) with a cladding, such as Zircalloy. Also see *Fission; Nuclear Reactor.*

FUEL USE ACT The Power Plant and Industrial Fuel Use Act, a law enacted by the United States Congress in 1978 which establishes the conditions for use of oil and natural gas as fuel in new industrial and electric utility installations. One of the conditions is that utilities must curtail the use of natural gas and oil for power generation by 1992. The Act encourages the use of coal for power plants.

FUGITIVE EMISSIONS Unintentional emissions into the air, such as dust blowing from a pile of tailings, coal, or shale.

FUMES Vapors rising from burning or vaporizing substances.

FUNDAMENTAL UNITS The primary dimensions of mass, M, length, L, and time, T, or Θ for time, with T for Temperature (Table F-5). The fundamental units are used to represent various physical and thermal relationships. Also see *Dimensions.*

FUNGI A class of plant spore-producing multicellular organisms which does not contain chlorophyll and which feed on organic matter.

FURNACE OIL A fraction of crude oil resulting from fractional distillation, between kerosene and light lubricating oil, used for fuel for furnace and boiler operation. Also see *Fractional Distillation.*

TABLE F-5
Fundamental Units

Basic units: M, mass; θ, time; L, length; T, temperature	
A	Area, L^2
c	Specific heat, heat capacity, L^2/θ^2T
c_a	Concentration, M/L^3
c_b	Specific vapor capacity, $L\theta^2/M$
c_m	Mass capacity, L^3/M
D, α	Diffusivity, L^2/θ
E	Entropy, ML^2/θ^2T
E_a	Activation energy, L^2/θ^2
F	Force, ML/θ^2
f	Frequency, $1/\theta$
g	Acceleration, L/θ^2
h	Heat transfer coefficient, $M/T\theta^3$
k	Thermal conductivity, $ML/T\theta^3$
k_c	Mass transfer coefficient, L/θ
L	Latent heat, L^2/θ^2, or characteristic dimension
M	Moisture constant
P, p	Pressure, $M/L\theta^2$
q	Heat flux, ML^2/θ^3
Q	Heat liberated/unit mass, L^2/θ^2
s	Humid heat, L^2/θ^2T
V	Velocity, L/θ
v	Volume, L^3
B	Coefficient of bulk expansion, $1/T$
u	Dynamic viscosity, $M/L\theta$
ρ	Density, M/L^3

FUSAIN A component in most coals which gives a smooth black appearance. It is brittle and high in ash, getting its name from fusite, the major constituent. Also see *Mother of Coal.*

FUSED SALT REACTOR A reactor type that would utilize a molten, fuel-bearing salt in the core. The advantage of this design is that it permits continuous fission product removal from the reactor. This type of reactor is still in the development phase and is not being pursued at present.

FUSION FUELS Fuels for the fusion reaction. Deuterium and tritium are the two fusion fuels.

FUSHUN PROCESS A process of retorting oil shale at temperatures of 430-540°C (800-1000°F) to produce oil from kerogen.

FUSION (nuclear) A thermal reaction during which two or more light atomic nuclei combine or fuse together to produce a heavier nucleus. Fusion of nuclei occurs only at very high temperatures. During nuclear fusion considerable energy is released. A device in which a controlled fusion process occurs to produce heat is called a thermonuclear reactor. One type of thermonuclear reactor uses pulsed lasers to heat tiny fuel pellets to extremely high temperatures to achieve fusion. The other type of system confines plasma material using magnetic fields and usually seeks to achieve high temperatures using ohmic heating of the plasma. Approximately 10^{14} W are needed for 10^{-9} sec to initiate a thermonuclear process. Also see *Fission*.

G

GADOLINIUM A rare earth element used as a neutron absorber in the nuclear industry. The atomic number of gadolinium is 64.

GAGE, GAUGE A device which gives a reading or value of a primary measuring element. Applied to a point of measurement, such as a pressure gauge.

GAGE PRESSURE The basis of representing fluid pressure in which atmospheric pressure is equivalent to zero, identified as zero gage pressure or 0 psig, pounds per square inch gage. At 1 atm pressure the absolute pressure is 101.325 kPa (or 101.325 kN/m^2) or 14.7 psia. Also see *Conversion Factors.*

GAL A unit of measure of gravity of 1 cm/sec^2, named after Galileo.

GALLATIN Also called dead oil; an oil made from coal tar.

GALLIUM ARSENIDE SOLAR CELL A device for changing solar energy to direct current electricity for which the active element is gallium arsenide (GaAs). The normal efficiency of conversion is from 13 to 19%, with a maximum theoretical efficiency of 27%. Also see *Solar Cell.*

GALLON A unit of volume used primarily for liquid measure having different values in the United States and England (Imperial). The U.S. gallon (liquid) is 231 in^3, 0.133 ft^3, or 3.7854×10^{-3} m^3, or 3.785 liter (L). The Imperial gallon is 4.55 ℓ or 4.546×10^{-3} m^3 or 278.3 in^3. The Imperial gallon is the volume of 10 lb of water at standard conditions. One U.S. gallon is 0.83 times the Imperial gallon. Also see *Liter.*

GALVANIC CELL A component of a battery which converts chemical energy directly to electrical energy. A galvanic cell is made up of two dissimilar materials separated from each other by an electrolyte.

GAMA Gas Appliance Manufacturers Association.

GAMMA Refers to gamma radiation and gamma ray (λ-ray), a high energy short wave electromagnetic radiation, similar to, but of shorter wavelength than X-ray. Gamma radiation may accompany the loss of alpha- or beta-particles but in itself is not a particle. The gamma rays penetrate deeply. Also see *Alpha; Beta; Electromagnetic; X-ray*.

GARBOIL An oily liquid produced from garbage and other solid wastes. The oil is used in oil-burning boilers and as a supplement to fossil-based oil.

GARRETT PYROLYSIS PROCESS A process for treating solid waste using low temperature pyrolysis to produce a viscous fuel oil with a heating value of 24,420 J/kg (10,500 Btu/lb) and char. The gas produced from the pyrolysis process is used for sustaining the process.

GAS, CALORIFIC VALUES See Calorific Values.

GAS, NATURAL See Natural Gas.

GAS COAL The name given to a coal, free of sulfur, which upon distillation yields a hydrocarbon vapor which can be used for illumination.

GAS CONSTANT The constant of proportionality in the law of ideal gases represented by R (Table G-1) in the equation

$$pV = mRT$$

which is the general gas law, where

p = absolute pressure
V = volume
m = mass
T = absolute temperature

GAS-COOLED FAST-BREEDER REACTOR (GCBR) A fast breeder reactor concept that uses a gas, e.g., helium, rather than a liquid metal as a coolant. It has not undergone nearly so much development in this country as the LMFBR. Gas-cooled reactors are suitable as power sources for gas turbine drives.

GAS-COOLED REACTOR A nuclear reactor which uses a gas as a coolant (Figure G-1). Gas-cooled reactors are the most common in other countries but not in the United States. Several characteristics give gas cooling advantages: Gases have low absorption cross sections;

TABLE G-1
Gas Constants

	R		Molecular weight,	Specific heat	
	J/kgK	ft lb/lb°R	M	c_p Btu/lb°R	c_v Btu/lb°R
Air	287	53.35	28.9	0.24	0.17
Carbon dioxide, CO_2	189	35.1	44	0.21	0.16
Carbon monoxide, CO	296	55.1	28	0.18	0.25
Hydrogen, H	4124	767	2.02	3.42	2.44
Nitrogen, N	296	55.1	28.02	0.247	0.176
Oxygen, O	260	48.3	32	0.217	0.155
Steam	462	85.8	18.02	0.41	0.36

Conversion factors: For R (specific) gas constant, 1 ft lb/lb°R = 5.38 J/kgK. For c_p and c_v, 1 Btu/lb°R = 4.188 kJ/kgK. Molecular weight in SI is called molecular mass.

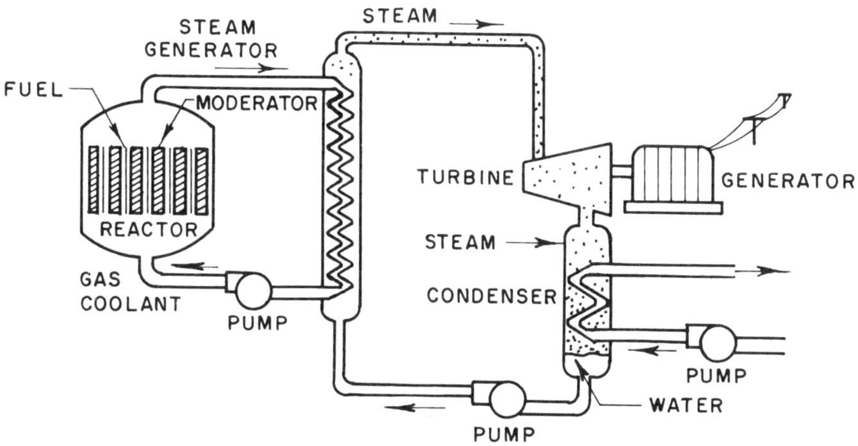

Fig. G-1 Gas-cooled reactor *(Schultz and Whicker, 1980)*.

they can be heated to a high temperature without pressurizing; and suitable gases, such as carbon dioxide, are available at low costs.

GAS CYCLING The procedure followed when drilling for oil in which the gas from the well is recovered and injected in the oil reservoir to assist in removal of the oil. Also see *Gas/Oil Ratio*.

GAS LIQUEFACTION (also written as Liquifaction) The propane and butane components of natural gas may be removed by compression and cooling, then liquefied and separated, making liquefied petroleum gas (LPG), compressed and placed in containers or tanks as a liquid and used for power, heating, cooling, and lighting. Natural gas, which is primarily methane, is usually piped as a gas to the point of consumption, but can be liquefied and is known as liquefied natural gas (LNG). See Methane; Natural Gas.

GAS OIL A fraction of crude oil removed during distillation. It is removed after kerosene, has a flash point of 168°F, is in the middle distillate range, which boils between 200 and 350°C. Gas oil may be converted to gasoline. Gas oil is used for stationary diesels.

GAS-OIL HYDROTREATER A cracking process used to remove sulfur and nitrogen from heavier crude stocks, making it acceptable for further processing, using heat treatment under pressure.

GAS SYNTHESIS The production of hydrogen and carbon monoxide used to make gas for fuel or other petrochemical feedstock.

GAS TURBINE A rotary heat engine in which the fuel is burned in a pressurized chamber and expanded to produce power to drive a shaft to a powered device or generator (Figure G-2). A gas turbine may be classed as open cycle in which the fuel is burned in compressed air and used for power, then exhausted, or as a closed cycle unit in which the heat is transferred via heat exchanger to a self-contained volume of gas which is recirculated. In the power industry, gas turbines are used for handling peaking loads to produce electricity, primarily because of their quick start-up.

GAS WATER The name given to water which has absorbed the gas present when passed through coal.

GASEOUS DIFFUSION A process used to separate the uranium isotopes ^{235}U and ^{238}U. Separation results from the difference in the diffusion rate of the gas, uranium hexafluoride, depending on whether the uranium in the molecule is ^{235}U or ^{238}U.

GASEOUS FUEL A general description of those fuels in a gaseous form, such as natural gas, and those made into a gaseous form from wood, coal, oil, etc. Also see *High-Btu Gas; Low-Btu Gas; Medium-Btu Gas*.

GASIFICATION The conversion of solids such as coal, lignite, coal tar, oil shale, and biomass and liquids such as oil, to a gas using appropriate temperature and pressure processes, without leaving a combustible residue. The formation of gas from carbon material to gaseous products which might include reaction with air, oxygen, steam, carbon dioxide with products CO, CO_2, H, and CH_4. Also see *Coal Gasification; Oil Gasification.*

GAS/OIL RATIO The amount of gas in the oil produced from a well at atmospheric conditions expressed in cubic feet per barrel. An average value is 1500 ft^3/bbl.

GASOHOL A fuel made of a blend of ethanol (10%) and gasoline (90%) for internal combustion engines. Also see *Alcogas; Alcohol; Diesohol; Ethanol; Methaline.*

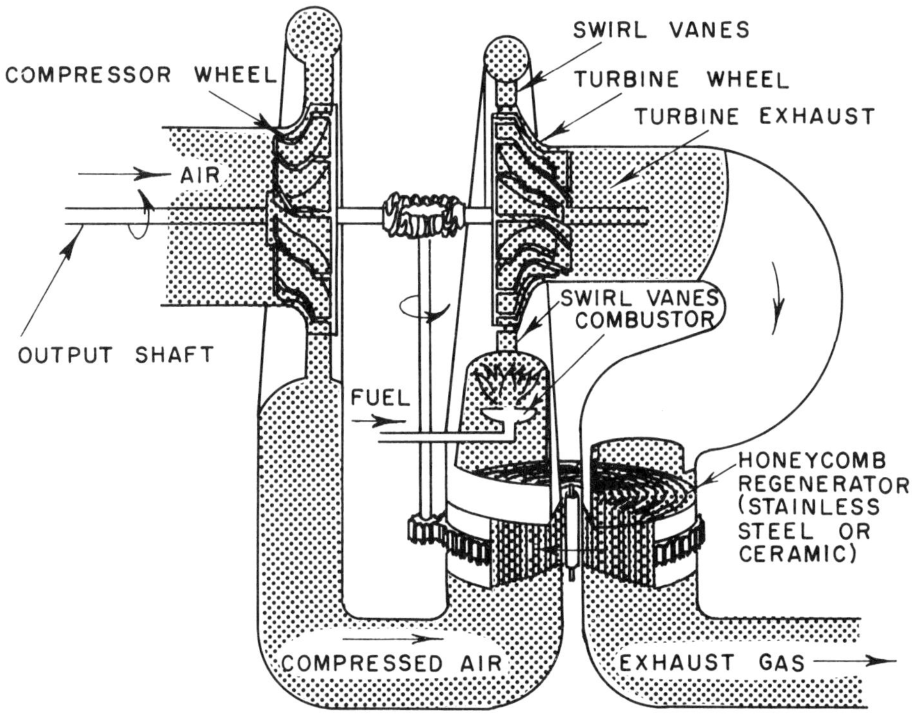

Fig. G-2 Gas turbine.

GASOLINE A colorless, volatile, liquid, inflammable automotive fuel in the C_5 to C_8 hydrocarbon range. Gasoline is the light distillate which is in the range of 30° to 200°C (86° to 392°F) which occurs in the fractional distillation of crude oil. Gasoline has a boiling point of 15° to 199°C (59° to 374°F). Gasoline is often called "gas."

GCBR Gas-cooled fast-breeder reactor.

GCFBR Gas-cooled fast-breeder reactor.

GCHWR Gas-cooled heavy-water-moderated reactor.

GCR Gas-cooled reactor.

GEGAS PROCESS A process of coal gasification developed by General Electric Company which produces low-Btu gas. Also see *Low-Btu Gas.*

GENERATION, ELECTRIC See Electricity Generation.

GENERIC SYSTEM A term used to describe the original basic unit, equipment, or process from which others have been derived. As an example, the Lurgi process could be considered generic for several other coal gasification systems.

-genic A suffix referring to generation of or production of.

GEOLOGIC TIME Long periods of time, represented by millions of years, used to represent time or events (Figure E-3). The era is the largest unit of time. The era consists of periods. The period is divided into epochs. The epoch is divided into ages. The present geologic time, the Cenozoic era, is 65 million years in length, and consists of three periods:

Pleistocene	1.0 million years to present
Quaternary	2.5 to 1.0 million years ago
Tertiary	65 to 62.5 million years ago

Also see *Cosmic Year; Epoch; Era; Period.*

GEOPRESSURE The pressure on geothermal fluids in the earth, approximately 20,000 psi.

GEOPRESSURIZED GAS Natural gas under great pressure in the earth and dissolved in liquid.

GEOSPHERE The solid portion of the earth's crust including the bodies of water associated with the surface of the earth. The lithosphere is the crust of the earth. Above it is the hydrosphere which includes the water bodies and water vapor. Together they make up the geosphere. The biosphere is the portion of the earth in which life exists and grows. Also see *Atmosphere; Biosphere.*

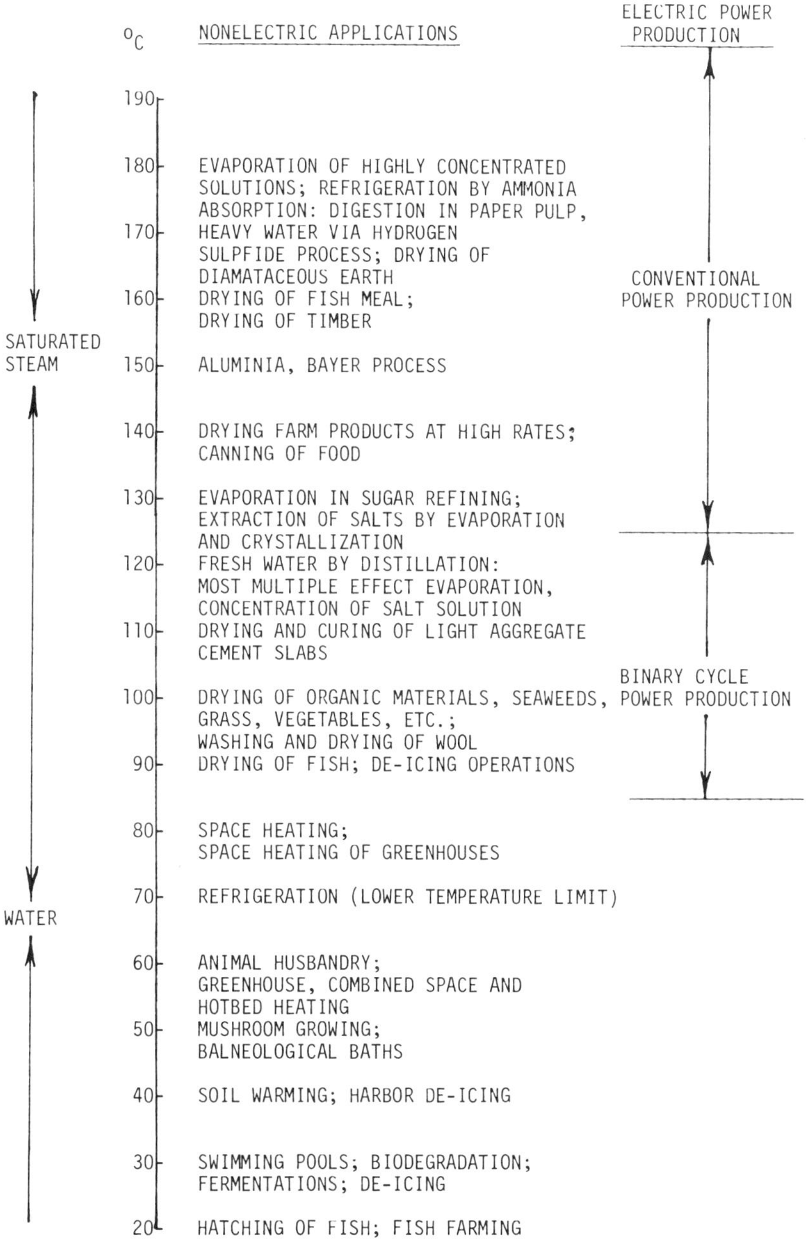

Fig. G-3 Geothermal energy for heating *(Lindal, 1973)*.

GEOTHERMAL ENERGY The energy in the earth's crust, which may be removed for use on the surface of the earth: Hot water at 60° to 120°C; steam (150° to 300°C); and hot rocks (300°C at 10,000 m depth). On the average, there is a geothermal gradient of 30°C/km, providing a low rate of heat transfer up through the surface of the earth equivalent to about 1/10,000 the rate from the solar system. The steam in the earth's crust may be used to generate electricity, drive steam turbines, provide hot water for space heating and processing (Figure G-3). Water or other fluids can be circulated through hot rocks to gain heat. From hot fluid cavities, dry steam or brine may be available. Some areas of the surface of the earth have a high temperature heat source beneath. The heat comes from radioactivity in the earth, movement of rocks in the earth, and chemical changes which occur, such as fossilization. Also see *Geothermal Gradient; Heat Flow, Earth; Magma; Mesothermal.*

GEOTHERMAL GRADIENT The heat source in the center of the earth provides a flow of heat through the earth's crust at an average flux over the face of the earth of 1.5 microcalories/cm^2s. The temperature difference with depth varies greatly over the face of the earth, being an average of 30°C per km depth in the earth's crust. Also see *Heat Flow, Earth.*

GIANT KELP Included in the brown algae, a large vine-like plant which grows up to 50 m (150 ft) in length attached to the sea bottom. The giant kelp is the largest in existence. Also see *Brown Algae; Kelp.*

GIBB'S ENERGY See Free Energy.

GIGA ELECTRON VOLTS (GeV) See Billion electron volts (BeV).

GILLETTE SYNDROME Named after the town of Gillette, Wyoming. It refers to the social impacts on boom towns that result from rapid energy development without proper planning.

GILSONITE A variety of natural asphalt which has a black color and brilliant luster, laced with brown streaks.

GNP Gross national product.

GOBAR The name used in parts of Asia for cow manure which is often used for production of biogas by anaerobic digestion. Also see *Biogas.*

GONDOLA CAR The most widely used type of railroad coal cars, usually of 90,910 kg (100 ton) capacity. The gondola car is of open top design, and may have drop ends and drop floors. There is also a solid bottom gondola which can be unloaded either by using clamshells or other scoops or by turning the complete car with a rotary device. Some gondola cars can be rotary dumped without uncoupling the cars. Also see *Coal Cars.*

GOR See Gas/Oil Ratio.

GPT, gpt Gallons per ton.

GRADES OF COAL A categorization of coal on the basis of constituents. The grade of coal is dependent on a combination of factors, such as sulfur content. The rank of coal depends on its age. Also see *Coal Classification; Rank.*

GRADIENT The rate of change of a variable quantity with respect to time or distance, such as °C/m or °F/ft and Δp/m, or for slope as m/m. The term also refers to the gradient of a function in vector notation, written as grad f or ∇f where

$$\nabla f = if_x + jf_y + hf_z$$

Also see *Geothermal Gradient.*

GRAM A unit of mass in the c-g-s system, with 453.59 grams per pound, 1000 grams per 1 kilogram, and 15.432 grains per 1 gram. In the SI units, the basic unit of mass is the kilogram (kg).

GRAM-CALORIE See Calorie.

GRAM MOLECULAR WEIGHT (GMW) The molecular weight of a compound stated in grams. Thus, methane, CH_4, has a gram molecular weight of 16 grams.

GRAND COULEE The name of one of the largest concrete dams with a head of 550 ft and the largest hydroelectric power plant in the world, located at Grand Coulee, in the north central part of the State of Washington. The generating capacity is 9771 megawatts. The Itaipu hydrofacility in Brazil-Paraguay, which is under construction, will have a larger production capacity. Also see *Dams; Hydroelectric.*

GRAPHITE A common form of carbon in which the atoms are arranged in layers, each layer consisting of a hexagonal pattern of atoms. Graphite, used initially for pencil leads, now is used as a constituent in lubricants, electrodes, and paints and as a moderator in some types of nuclear reactors. Graphite is also known as plumbago and blacklead. Also see *Carbon; Moderator.*

GRAVITATIONAL ENERGY A form of energy associated with gravitational fields. Gravitational energy can be released in other forms as a result of accretion of matter such as occurs in condensation of stars and galaxies. One theory is that the thermal energy in the earth's core was supplied from gravity upon collapse of the gas from which the earth derived.

GRAVITY The attractive force by which a body is pulled toward the center of the earth. The attractive force between two bodies is proportional to the product of their masses and inversely proportional to the square of the distance between them.

GRAVITY WAVES ENERGY The energy available from the deep sea and ocean waves which are propagated by gravitational forces on the water. Wave motors have been developed to capture the energy of the waves. Also see *Ocean Thermal Energy Conversion.*

GREENHOUSE EFFECT A term describing the property that the atmosphere of the earth permits short wave radiation from the sun to pass through it to the earth's surface but absorbs the long wave radiation emitted at the earth's surface. The action of the atmosphere is one of moderating temperature changes and keeping the surface of the earth within a temperature range to permit organisms to live.

GREY (Gy) A unit of absorbed radiation equal to 100 rad. See Rad.

GRI Gas Research Institute, 8600 West Bryn Mawr Avenue, Chicago, IL 60631.

GRID The name applied to the interconnected parts of the electric transmission system. Also see *Power Wheeling.*

GRINDING The reduction in size of material with a mill or pulverizer to improve materials handling and uniformity of size, and to increase surface area for subsequent processing, drying, combustion, or reaction. Grindability index of coal is based on a standard coal of 100, and is above 100 for harder coals and below 100 for softer coals. (ASTM Standard D409.)

GROSS ENERGY The quantity of heat, in calories, released when a substance is completely oxidized in a bomb calorimeter containing 25 to 30 atmospheres of oxygen. The term heat of combustion is also used. Also see *Heat of Combustion.*

The total amount of energy involved to point of use or delivery, including all losses. The gross energy is always larger in value than net energy for a particular process. Also see *Net Energy.*

GROSS ENERGY, FOOD The gross energy of food consumed which is used by a person or animal. An example of the energy utilization and heat production of a lactating cow is shown in Figure G-4.

GROSS ENERGY REQUIREMENTS (GER) The total amount of primary energy needed to support a process, produce a commodity, or carry out an operation. Thus, although 3412 Btu equals 1 kWh, 10,000 Btu of primary fuel are required in a typical thermal power plant to produce 1 kWh of electricity. Also see *Primary Energy Equivalent.*

GROSS HEATING VALUE See High Heating Value.

GROUND STATE The lowest energy quantum state in which a system

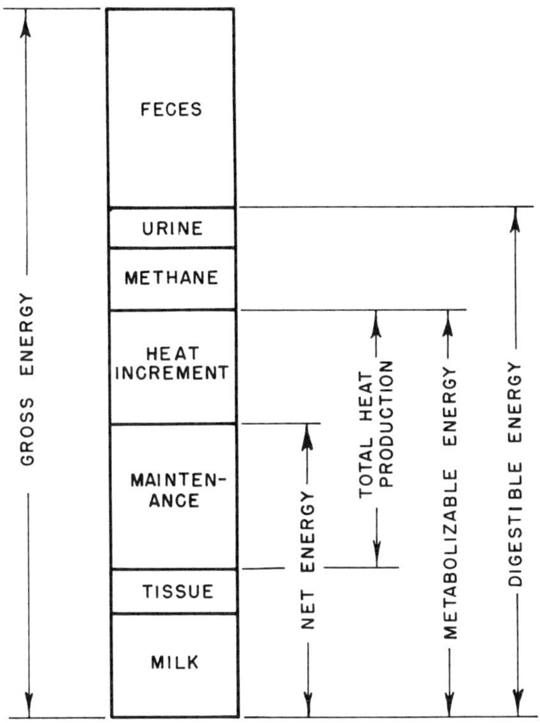

Fig. G-4 Gross energy, food.

can exist. A system such as an atom, molecule, crystal, etc., can normally be in any of many possible quantized energy states. The lowest is called the ground state.

GULF CCL PROCESS A process for producing light oil and heavy oil from a slurry of coal in a hydrogen atmosphere.

G-VALUE The value of the acceleration due to gravity. This value varies slightly from place to place on the earth's surface. In fact, measuring g is a mineral prospecting technique.

GW Gigawatt.

GY See Grey.

H

H-ABSTRACTORS Hydrocarbons with unsaturated carbon which can use the hydrogen abstracted from coal.

H-ACCEPTORS Hydrocarbons which have unsaturated carbon which accept the hydrogen from the coal.

H-COAL A liquefaction process of producing synthetic crude oil from coal. The pulverized slurry of coal is treated at a pressure of 200 atmospheres and temperature of 450°C (842°F). A pelletized metal catalyst is used to accomplish catalytic hydrogenation. The synthetic crude oil can be refined to make middle and high distillates. Also see *Exxon Donor Process (EDP); Hydrogenation*.

H-DONORS Partially hydrogenated aromatic hydrocarbons which provide hydrogen to dissolved coal to prevent char formation. Also see *Coal Liquefaction; Hydrogenation*.

H-SHUTTLERS Process in which hydrogen, from the coal or SRC, transfer is a mechanism for dissolving coal, basically by shuttling (transferring) hydrogen from one position to another in the coal.

HAFNIUM A metallic element, Hf, used as a neutron absorber in the nuclear industry. Hafnium is the active neutron absorber in the control rods of some types of nuclear reactors. The atomic number of hafnium is 72.

HALF-LIFE A term used to represent the time at which a radioactive substance has decayed to one-half of the original material, activity, or strength (Table H-1). The relationship has been applied to many physical, chemical, and biological phenomena. Also see *Life*.

HALF-THICKNESS (nuclear) The thickness of material which will absorb half the radiation impinging on it. This value depends on

TABLE H-1
Half-Life of Some Isotopes

Isotope		Approximate half-life, yr
Carbon	^{14}C	5,568
Krypton	^{85}Kr	10.7
Plutonium	^{239}Pu	24,100
Strontium	^{90}Sr	25
Thorium	^{232}Th	1.4×10^{10}
Tritium	^{3}H	12.26
Uranium	^{235}U	7.1×10^{8}
Uranium	^{238}U	4.5×10^{9}

the type and energy of the impinging radiation as well as the absorber used.

HALOGENS The members of the group VIIb in the periodic table of elements, bromine, chlorine, fluorine, iodine, and a less known element, astatine. These elements are non-metals which are involved in many energy-related matters, such as activity with oxygen and hydrogen, corrosion of metals, impurities in minerals, acids, salts, etc.

HAMMER MILL A type of grinder used to comminute products which may be used preceding processing to provide a more uniform small size of product and to increase the surface area. The hammer mill is used to prepare solid fuels for combustion. Energy added during grinding and with air flow may appreciably reduce moisture content of the material. Also see *Grinder.*

HARD COAL In the United States hard coal is anthracite coal. In Great Britain hard coal is bituminous and anthracite coal, or those coals above lignite. Also see *Anthracite Coal; Bituminous Coal; Lignite; Malting Coal.*

HARD RADIATION Radiation that penetrates large thicknesses of matter. The term "hard" is often applied to high frequency X-rays and gamma rays. Hard gamma rays are the most difficult radiations to absorb in the shielding materials placed around radioactive sources.

HARDNESS The resistance of a material to crushing, indenting, or abrading when impacted, measured on a Moh scale of 1 to 10, soft to hard, or as given by a time-force-physical change plot. In general, products become harder upon drying. Hardness may be used as a means of rapidly determining indirectly the moisture content. For metals, the Brinnell scale represents the hardness.

Hardness also refers to water containing the salts of calcium.

HDR Hot dry rock.

HEAD A term for representing the equivalent of pressure exerted in terms of millimeters, centimeters, meters, inches, or feet of water or mercury (Hg). Also see *Pressure*.

HEAT A form of thermal energy which flows from a hot to a cold body with the flow dependent on the temperature difference. A change in thermal energy may be associated with calories, Btu, temperature, radiation, change of state, expansion or contraction, and electrical effects. Also see *Heat Transfer*.

HEAT, LATENT See Latent Heat.

HEAT, RADIANT See Radiation; Electromagnetic Radiation.

HEAT, SPECIFIC See Specific Heat.

HEAT CAPACITY The quantity of heat necessary to raise the temperature of a body or system by one degree, either in calories/°C or Btu/°F.

Heat capacity, C, is defined as C = dU/dT where U is the change of internal energy of the system. The heat capacity of one gram of water is 1.00803 cal/°C at 0°C at 1 atm. For a gas the heat capacity at constant pressure is larger than heat capacity at constant volume. The specific heat is the heat capacity of a unit amount of material. Also see *Specific Heat*.

HEAT CONDUCTIVITY See Conduction, Heat; Thermal Conduction.

HEAT CONTENT See Heat Capacity.

HEAT EXCHANGER A device used to transfer heat from one fluid to another. Also see *Indirect Heater*.

HEAT FLOW The quantity of heat flow per unit time, per unit cross-sectional area. Units and conversions are: 1 Calorie/sec cm^2 = 4.186 watts/cm^2 = 3600 Calories/hr cm^2 = 13,272 Btu/hr ft^2 = 318,531 Btu/day ft^2 = 41,873 W/m^2.

HEAT FLOW, EARTH Heat from the hot inner section of the earth moves to the crust and surface of the earth, primarily by conduction. The average heat flow over the earth averages about 49.42 kJ/m^2s, less than one-ten thousandth the flow of heat onto the earth from the sun. Also see *Geothermal Energy; Geothermal Gradient*.

HEAT FLUX DENSITY The rate of thermal energy flow per unit area. Usual units are J/m^2s, W/m^2, cal/cm^2s, and Btu/ft^2hr. Also see *Energy Density; Specific Energy*.

HEAT OF COMBUSTION The quantity of heat produced by burning a certain quantity of material, expressed in calories/gram molecular weight, Kilogram-calories/mole; Cal/gram, cal/gram, Btu/lb, or kJ/kg. Also see *Calorific Value; Combustion; Fuel; High Heating Value; Low Heating Value; Wood Combustion.*

HEAT PIPE A heat pipe provides a means of recovering heat from an external source (Figure H-1). A heat exchanger pipe transports heat by evaporation and condensation of a liquid in an enclosed pipe.

A heat pipe is used to support the Alaska pipeline on the permafrost. The foot of the heat pipe, which is kept cold, rests on the frozen land. Heat is dissipated from the top.

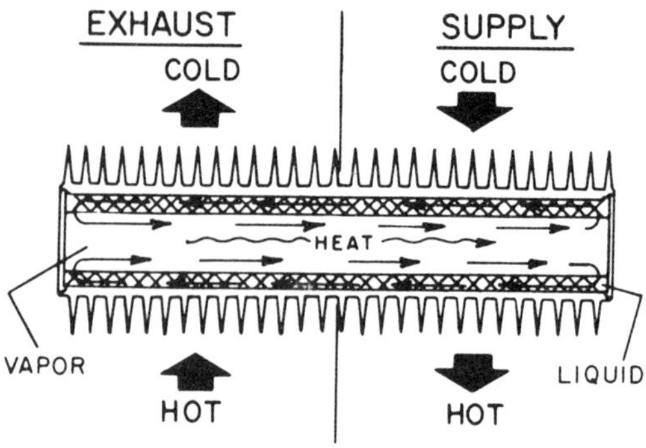

Fig. H-1 Heat pipe.

HEAT PUMP A device which extracts heat from a low temperature source and provides the heat to a higher temperature medium. The heat pump is similar to a refrigeration cycle, with the difference being that the condenser becomes the heat rejection device for heating (Figure H-2). The amount of energy in the form of electricity put into the system to provide heat is one-half to one-third the amount of heat which would be provided for direct resistance heating, giving a coefficient of performance (COP) of from two to three. Also see *Coefficient of Performance.*

HEAT RECOVERY The procedure of recovering and utilizing sensible and latent heat that would otherwise be wasted or rejected. Heat recovery procedures must consider the cost versus the benefit or return. Heat recovery is normally accomplished through direct or in-

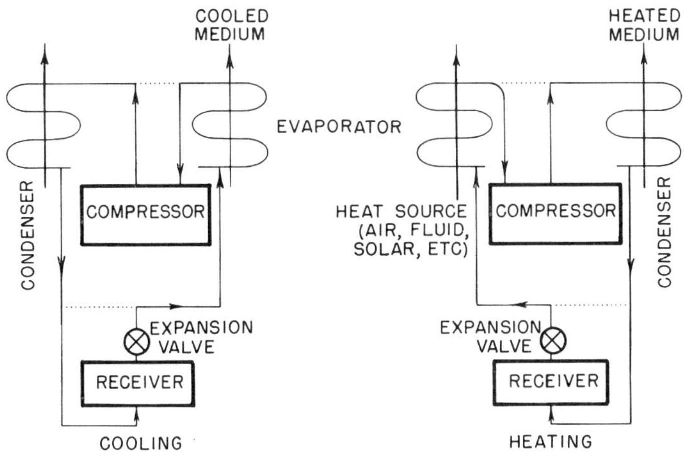

Fig. H-2 Heat pump cycle.

direct heat exchangers, heat pump, heat storage, heat wheel, waste heat boiler. Also see *Heat Storage; Heat Wheel, Waste Heat Boiler.*

HEAT STORAGE Storing and holding excess heat, usually above atmospheric temperature, for use at a later time. For example, solar energy after collection can be stored in rocks, chemicals (as sensible or latent heat), or water. Secondary energy storage can be as electricity in batteries, hydraulic energy in reservoirs, liquids under pressure, or mechanical energy in flywheels.

HEAT TRANSFER The movement of thermal energy through a body or across a space as a result of temperature difference. Transfer can take place by one or a combination of the three basic modes of heat transfer: (1) conduction (molecular vibration depending on the temperature), (2) convection (transfer of a fluid mass from one to another location depending on density), and (3) radiation (electromagnetic, depending on the surface characteristics and the absolute temperature). Also see *Conduction; Convection; Radiation.*

HEAT VALUE See Calorific Value; High Heating Value; Low Heating Value.

HEAT WHEEL A device for recovery of heat, also known as a Ljungstrom regenerator. The device consists of a wheel or rotor which turns slowly (1 to 3 rpm) in a housing made of two compartments. One compartment is for the cold air being heated; the other is for the hot air being cooled. In one revolution the rotating wheel is heated by the hot air or gas and gives up heat to the incoming cold air or gas. The wheel may rotate in a vertical or horizontal plane. The wheel has a large heating surface formed by plates, tubes, or cells.

HEATING VALUE See Calorific Value; High Heating Value; Low Heating Value.

HEAVY HEAVY HYDROGEN See Tritium.

HEAVY HYDROGEN See Deuterium.

HEAVY ION REACTIONS Reactions between target atoms and ion beams containing ions heavier than helium.

HEAVY WATER Deuterium oxide, D_2O, a compound of deuterium and oxygen which is used in nuclear reactors for neutron moderation or coolant. D_2O is separated from water using distillation and electrolysis processes. The physical properties of heavy water are: Boiling point 101.6°C (214.9°F); maximum density, 1,106 kg/l at 11°C (52°F); and melting point, 3.75°C (38.8°F). Also see *Light Water; Pressurized Heavy Water Reactor.*

HELIOSTAT A device which follows the sun to direct its rays onto a solar collector or mirror. Also see *Nutation.*

HELIUM An inert light gas, He, which is odorless, colorless, and not flammable (non-inflammable). Helium has an atomic number of 2, an atomic weight of 4, density of 0.1785 kg/m^3 at STP. Helium can be liquefied at 4.2 K (-269°C). Pressurized helium is used as a cooling agent in some nuclear reactors. Also see *High Temperature Gas-Cooled Reactor; K.*

HEMICELLULOSE One of the plant cellulose materials $(C_6H_{10}O_5)_n$ + $(C_5H_8O_4)_n$, which is a carbohydrate that cements the plant cellulose fibers together. Also see *Cellulose; Holocellulose.*

HEPA High-efficiency particulate air (filter).

HERTZ A term for designating frequency, one cycle per second, abbreviated Hz. Sixty cycles per sec equals 60 Hz.

HETEROCYCLIC The arrangement of atoms of a compound in a ring in which one or more of the atoms is different than the others; for example, one atom of the ring is not carbon. Nitrogen, oxygen, and sulfur atoms are most commonly in the ring of heterocyclic compounds. Also see *Homocyclic.*

HEXANE A light, volatile, flammable hydrocarbon liquid (C_6H_{14}), made from petroleum, forming part of gasoline. Hexane has a specific gravity of 0.66, boiling point of 69°C (156°F), and melting point of -94°C (-132°F). Also see *Gasoline.*

HEXENE See Hexylene.

HEXYLENE A liquid hydrocarbon, C_6H_{12}, also called hexene, obtained from petroleum. It has a density of 0.6732, boiling point of 63.5°C (146°F), and melting point of -98.5°C (-145.3°F).

HHV High heating value.

HIGH-BTU GAS A fuel gas containing at least 700 Btu/ft^3 (26 MJ/m^3). The major high-Btu gas is natural gas which has from 900 to 1100 Btu/ft^3, usually considered at 1000 Btu/ft^3 for energy studies. Also see *Low-Btu Gas; Medium-Btu Gas*.

HIGH FREQUENCY CURRENT (HFC) An abbreviation for high frequency current, which for electricity is a frequency between (3 and 30) x 10^6 cycles per sec. Also see *Radio Frequency*.

HIGH GRADE HEAT A term generally applied to heat at 212°F and above. Also see *Low Grade Heat*.

HIGH HEATING VALUE (HHV) The high heating value, also known as the gross heating value, is the amount of heat produced from burning a fuel, in J/kg or Btu/lb (Table H-2). The high heating value is obtained in a calorimeter by assuming that energy is removed from the combustion products until the water produced has been liquefied. The complete combustion of fuel including carbon, hydrogen, and sulfur in the presence of adequate oxygen gives carbon dioxide, water, and sulfur dioxide. Also see *Low Heating Value*.

TABLE H-2
High Heating Values, Typical

	Heating values	
	Btu/lb	MJ/kg
Butane	21,293	49.4
Cane sugar bagasse	7,900-9,100	18.3-21.1
Carbon	14,086	32.7
Carbon monoxide	4,343	10.1
Cotton ginning waste	7,900	18.5
Fuel oil	18,500	42.9
Gasoline	20,200	46.9
Hardwood, dry	8,100-8,900	18.8-20.6
Hydrogen	60,950	141.5
Kerosene	19,900	46.2
Methane	23,860	55.5
Propane gas	21,650	50.2
Softwood	8,800-9,800	20.4-22.7
Straw, 15% moisture	5,400	12.5
Wood, air dry	5,400-6,800	12.5-15.8

HIGH LEVEL RADIOACTIVE WASTE (HLW) Waste consisting of concentrated fission products and other waste products produced during nuclear reactor operation. The waste is characterized by high levels of radiation, long half-life, and high heat generation. Although there are apparently 270,000 m^3 of high level waste stored in the United States, only 2,300 m^3 have been generated as a result of commercial reprocessing, the remainder resulting from military operations. *(Gloyna, 1980.)*

HIGH RANK COAL Coal in which there is more than 84% carbon on an ash-free basis, and not more than 4% moisture. Also see *Coal Classification; Coal Rank.*

HIGH SULFUR COAL Coal in which the amount of sulfur based on heat content of the coal is more than 1.68 lb of sulfur for each 10^6 Btu of heat content.

HIGH TEMPERATURE GAS-COOLED REACTOR (HTR) A type of reactor for production of electricity which is helium cooled and uses enriched uranium carbide for the fuel. Also see *Fission Reactor; Nuclear Reactor.*

HIGH VOLATILE COAL Bituminous coal which has more than 31% volatile matter and less than 69% free carbon on a dry (mmf) basis with at least 14,000 Btu/lb moist (mmf). Additional classifications can include high volatile bituminous B and C coals. Also see *Bituminous Coal; Coal Classification; Coal Rank.*

HITEC A nitrate eutectic compound used for storage and transport of thermal energy in an active solar energy heating system.

HLW High-level radioactive wastes.

HOC Heavy oil cracking.

HOGGED FUEL A fuel consisting of the waste wood from lumber or wood pulp operations, including bark. Also see *Bark.*

HOLOCELLULOSE The name describing the combination of cellulose and hemicellulose in wood. Also see *Cellulose; Hemicellulose; Wood.*

HOMOCYCLIC The arrangement of atoms of a compound in a ring in which all of the atoms are the same, for example, carbon. If the atoms vary, the compound is heterocyclic. Also see *Heterocyclic.*

HPFL High performance fuels laboratory.

HORSEPOWER A measure of the rate of doing work, which is 33,000 ft lb/min for 1 horsepower, or 746 watts. The useful work which can be performed by a power device is called brake horsepower.

HOT ATOM (nuclear) A radioactive atom.

HOT CELL (nuclear) A chamber in which work is carried out on highly radioactive materials. The typical hot cell is shielded on all sides to protect personnel outside the cell from the radiations emitted by the radioactive materials within it. At least one side of the cell is made of layers of glass or plastic which permit the interior to be viewed but which still are thick enough to provide the shielding required. The hot cell also protects workers from airborne radioactivity and confines spills to the cell volume. Work inside the cell is done using remotely controlled mechanical "hands" or, in some cases, sealed glove protrusions.

HOT DRY ROCK (HDR) The geothermal resource base of the dry hot rock in the crust of the earth down to a depth of 10 km and at temperatures above 150°C. The total geothermal sources are estimated to be 13.2×10^6 quads, of which 99% exists in hot dry rock.

HOT ROCKS In geothermal energy, the high temperature rocks in the earth's crust, a potential source for thermal energy. Also see *Geothermal; Heat Flow, Earth; Hot Dry Rock.*

HTGCR High-temperature gas-cooled reactor.

HTPB Hydroxy-terminated polybutadiene.

HTGR High-temperature gas-cooled reactor.

HTR High-temperature gas-cooled reactor.

HUMACITE A general name for bitumens varying from hard to elastic. In physical composition it is an emulsion of humic acid hydrocarbons with water, not soluble in organic solvents.

HUMIC Carbonaceous material with 4-6% hydrogen in soils and sediments derived from plants. Refers to a class of bituminous coals in which there has been limited decay. Also see *Bituminous Coal.*

HUMID HEAT The quantity of heat required to raise the temperature of one unit mass of dry air plus the water vapor which accompanies it by one degree. Expressed in terms of kJ/kgK, Btu/lb dry air-°F or kcal/kg dry air-°C.

HUMID VOLUME The volume occupied by one unit mass of dry air plus the water vapor it contains, expressed as ft^3/lb dry air or m^3/kg dry air.

HVAC Heating, ventilating and air conditioning.

HVAC High voltage alternating current system.

HVDC High voltage direct current system.

HWLWR Heavy-water-moderated boiling light-water-cooled reactor.

HWR Heavy water reactor.

HYBRID SOLAR SYSTEM For building heating and/or cooling, a system in which both active and passive approaches (for using solar energy) are incorporated. The term also is used for a hybrid solar and fossil energy supply to furnish energy to a heat engine to produce power (Figure H-3). Also see *Active Solar; Passive Solar.*

HYCRACKING A process of producing a high-quality gasoline naphtha from petroleum in a fixed bed using steam for catalytic hydrocracking. Also see *Catalytic Hydrocracking.*

HYDRANE GASIFICATION A process to convert coal to high-Btu gas of 900-1000 Btu/ft^3. Pulverized coal is heated to 750°C (1382°F) at a pressure of 70 atmospheres. Hydrogen is used to make a fluidized bed of char, thus providing gas from the fluid bed at 950°C (1742°F). The gas is 90% methane and the yield is 0.6 m^3/kg of methane from dry coal. Also see *Coal Gasification; Gasification; Kellogg Process; Koppers-Totzek.*

HYDRAULIC TURBINES Devices used to transfer the energy in flowing water into rotating mechanical power for driving a mill, machine, pump, electric generator, etc. These units convert 85% of the hydraulic to mechanical energy. There are three major designs: Pelton, Francis, and Kaplan turbine.

The Pelton is an impulse wheel in which a jet of water strikes specially designed buckets on the periphery of the wheel. The Francis is a reaction turbine in which water flows along the radius and leaves along the axis. The Kaplan turbine is of a propeller-type with variable pitch in which flowing water turns the propeller. Those turbines with propeller blades at an angle of 45° are diagonal-flow or Dariaz turbines. Also see *Bulb-Turbine; Dariaz; Francis Turbine; Kaplan Turbine; Pelton Wheel.*

HYDRIDE A metallic compound, such as iron-titanium or magnesium hydride, which, under pressure, can be used to store hydrogen molecules. The hydrogen forms chemical bonds on the lattice structure. Energy such as supplied by heating is required to release the hydrogen. Magnesium hydrides are now the most attractive means of storing hydrogen for vehicles, and require about 30% of the energy to release the hydrogen.

HYDRO Prefix meaning water, for example, hydroelectric generator, means water-powered generator; also used for hydrogen, or for water production, as a compound of hydrogen.

HYDROCARBON MINERAL A naturally occurring solid primarily composed of carbon and hydrogen.

HYDROCARBON SERIES Hydrocarbons are divided into various series. The four that comprise most of the naturally occurring petroleums

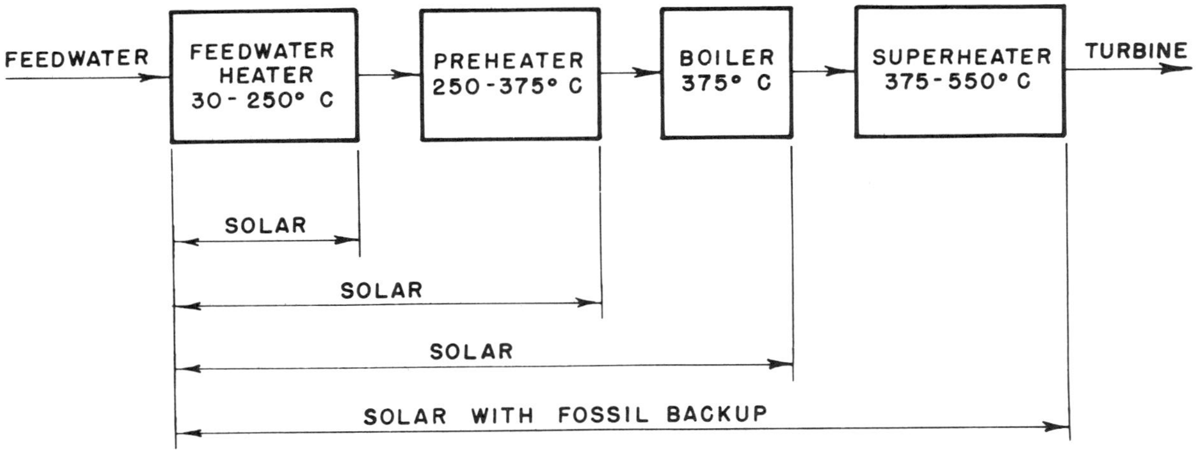

Fig. H-3 Hybrid solar energy and fossil energy system.

are the normal paraffin (alkane) series, the isoparaffin series (branched-chain paraffins), the naphthene (cycloparaffin) series, and the aromatic (benzene) series. Crude oils are referred to as paraffinic-base, naphthenic-base, or mixed-base (naphthenic-paraffinic) oils. Also see *Aromatic Series; Isoparaffin Series; Naphthene Series; Olefinic Hydrocarbons; Paraffin Series.*

HYDROCARBONS Organic compounds consisting primarily of carbon and hydrogen, the principal components of fossil fuels; some of the principal products produced from biomass such as methane. Some plants produce substantial amounts of hydrocarbons. Coal is a hydrocarbon mineral which contains carbon and hydrogen along with some oxygen compounds. Technically, natural gas and oil are the principal hydrocarbons occurring in nature. The densities, boiling points, and freezing points increase as the molecular weight of hydrocarbons increase. The smallest hydrocarbon molecules are gaseous while the largest are solids.

HYDROCRACKER A cracking device/process which usually follows hydrotreating of oil, which is more severe than hydrotreating, using higher temperatures and pressures and more hydrogen to break down heavier feedstocks.

HYDROCRACKING Usually used as an adjunct to catalytic cracking, a process in which hydrogen is added to materials with large molecules, followed by cracking at pressures of 70 to 150 atmospheres and moderate temperatures from 260 to 400°C (500 to 750°F). Also see *Fluid Hydrocracking; Hydropyrolysis.*

HYDROELECTRICITY Electricity produced by falling or flowing water (Figure H-4). Also see *Pumped Storage.*

HYDROFORMING An important process involved in the production of gasoline in which low-octane hydrocarbons are converted to higher octane fuel. Conversion is accomplished by heating to a high temperature under pressure in the presence of a catalyst.

HYDROFRACTURING The use of water or water solutions to break up minerals, rock or soil covering materials, or to fracture underground materials to release crude petroleum.

HYDROGASIFICATION A process which involves the addition of hydrogen to products of primary gasification to optimize the formation of methane, usually as a part of a heating process.

HYDROGEN A colorless, odorless gas, represented by H, which normally exists as H_2. The hydrogen atom is the lightest of all atoms, consisting of one proton and one electron. Hydrogen is an abundant element on the earth but usually combines to form compounds. The properties of hydrogen are: Atomic number is one; the molecular weight is 2.016; boiling point at 1 atm, -252.8°C (-423°F); and gross heat of combustion, 141,750 kJ/kg (61,100 Btu/lb) and 325 Btu/ft^3.

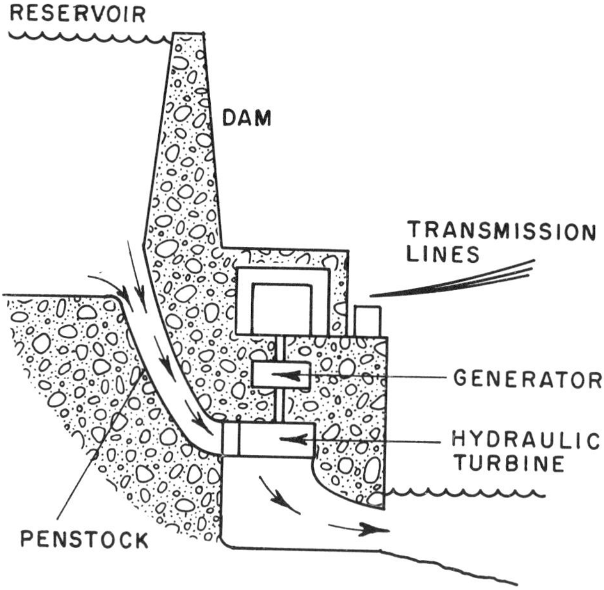

Fig. H-4 Hydroelectric generating facility *(Duderstadt, 1979)*.

The density of hydrogen gas is 0.0899 g/liter at STP. The critical temperature is -240°C (-400°F) which is at 12.8 atm, with the liquid at 31.2 g/liter. Hydrogen is explosive in mixtures of air of 4 to 74%. Hydrogen is receiving considerable attention as a secondary fuel or energy carrier. Hydrogen can be produced from water and hydrocarbons. Hydrogen-oxygen can be used in a fuel cell. The many possible sources and uses of hydrogen suggest a hydrogen economy (Figure H-5). Also see *Elements; Explosion; Fuel Cell.*

HYDROGEN LIQUEFACTION The process of cooling hydrogen gas to -253°C (-423°F) to form a liquid. The liquid is stored in vacuum-insulated containers. About 25% of the calorific or heating value of hydrogen is needed to liquefy the gas.

HYDROGEN STORAGE Hydrogen may be produced, usually on a continuous uniform rate, stored, then used from storage to meet peak loads or uneven demands. Hydrogen may be stored as a gas under high pressure, as a liquid at cryogenic temperatures, below -253°C (-423°F) and in metal hydrides. Also see *Hydride (Metal Hydrides)*.

HYDROGEN SULFIDE A compound present in some crude oil and gas, in varying quantities, usually considered an undesirable component.

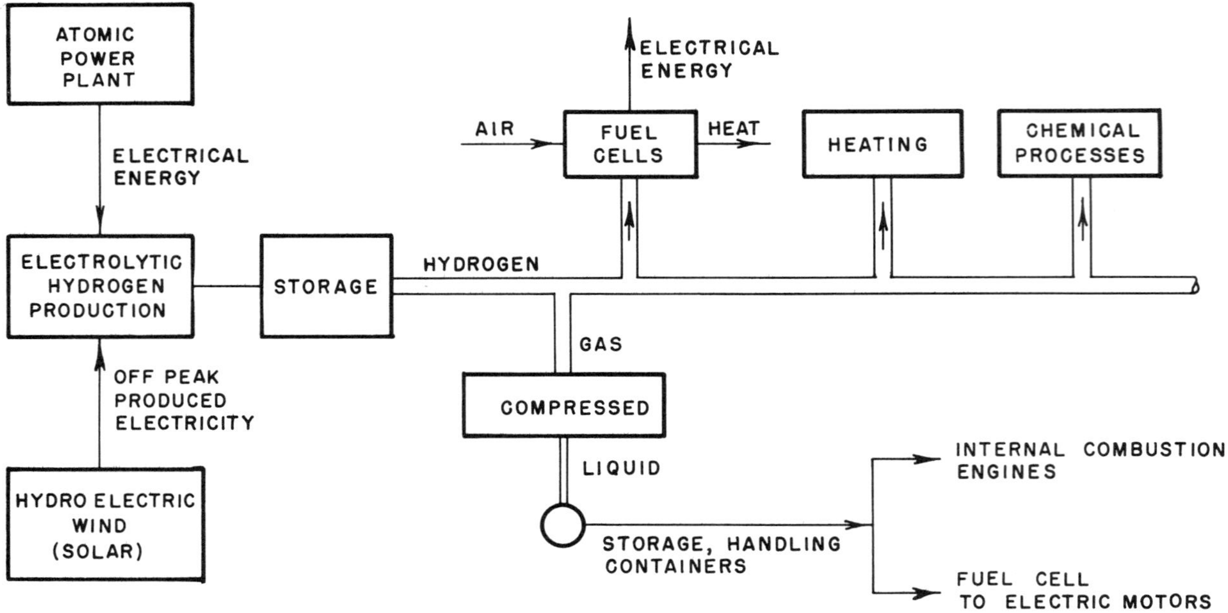

Fig. H-5   Hydrogen economy.

Hydrogen sulfide causes corrosion of equipment for storing, handling, and combustion and causes pollution of environment when burned. It is more economical in most cases to remove hydrogen sulfide prior to use as a fuel than to counter the negative effects of corrosion and pollution. Gas and oil with large quantities of hydrogen sulfide can be major sources of sulfur. Also see *Acid Gas; Claus Process; Sour Gas; Sour Oil.*

HYDROGEN/CARBON RATIO The relationship of hydrogen to carbon usually as atomic ratio, such that H/C for methane, CH_4, is 4 (Table H-3).

TABLE H-3
Hydrogen/Carbon Mole Ratios

Anthracite coal; coke	0.5
SRC	0.8
Coals	0.6 to 1.0
Synthoil; H-coal	0.8 to 1.2
Peat	1.0 to 1.4
Asphalt; tar sand	1.3 to 1.6
Shale oil	1.6
Heavy petroleum	1.5 to 1.7
Light petroleum	1.7 to 1.9
Distillate	1.8 to 2.0
Methane	4.0

Also see *Carbon/Hydrogen Ratio.*

HYDROGENATION The process of adding hydrogen to increase the hydrogen/carbon ratio of coal as it normally exists, and as is done in some coal gasification and coal liquefaction processes. Hydrogen is added as a gas in the presence of a catalyst at high temperatures and often at high pressures. The hydrogen reacts with certain organic compounds to saturate double and triple bonds. Also see *Coal Liquefaction; Extraction-Hydrogenation; Hydrogen/Carbon Ratio.*

HYDROGENOLYSIS The name given to the general process in which maximum quantities of gases are produced with a minimum of liquids as a result of hydrogen being added under pressure.

HYDROLIQUEFACTION A general term applied to coal liquefaction processes in which hydrogen under pressure is supplied to dissolve the organic components of the coal. Also see *Solvent Refining.*

HYDROLOGICAL CYCLE The circulation of water over and through the lithosphere (crust of earth), hydrosphere (water over the surface of the earth) and through the atmosphere (liquid/vapor). Solar energy provides the driving energy to cause the cycle to go providing energy for evaporation, convection, and condensation (Figure H-6). Also see *Advection; Atmosphere; Convection; Hydrosphere; Lithosphere.*

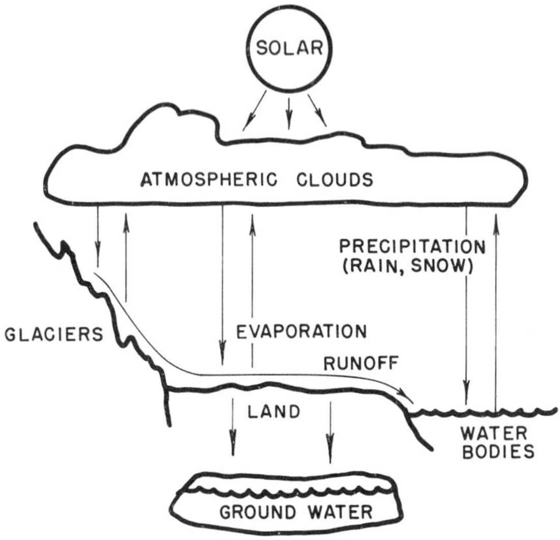

Fig. H-6 Hydrological cycle.

HYDROLYSIS A chemical process that breaks complex organic molecules into simple molecules. For example, starch and cellulose are hydrolyzed by water or by water with an acid or base or enzymes to produce simple sugars which can be fermented to produce alcohol.

HYDROPROCESSING Another name for solvent refining.

HYDROPYROLYSIS A crude oil cracking process in which the operation is done in the presence of hydrogen under pressure. As compared to hydrocracking, hydropyrolysis is done without catalysts, at higher temperature, and for a shorter time. Also see *Hydrocracking.*

HYDROREFINING Another name for solvent refining.

HYDROSPHERE See Geosphere.

HYDROTHERMAL Heat in water in the cavity or porous sections of the earth's crust which could provide a source of thermal energy. Also see *Geothermal; Hot Rocks.*

HYDROTREATING A general term applied to the process of adding hydrogen to a reaction, such as to oil produced by pyrolysis, or from coal to remove sulfur as hydrogen sulfide (H_2S) to provide a clean liquid fuel.

HYDROUS ALCOHOL Alcohol which contains more than 5% water, also called watery alcohol, usually 70-95% alcohol, which can be used for fuel.

HYDROXYL The radical OH^- which carries a negative charge. The hydroxyl ion is a part of all base reactions. A base substance containing an OH^- radical is a hydroxide.

HYGAS A gasification process for coal and lignite involving two steps: The coal or lignite is subjected to hydrogen-rich hot gas at 650°C (1200°F) to form methane, with the char formed in the first stage used to supply heat in the second stage (Figure H-7). Also see *Coal Gasification; Gasification.*

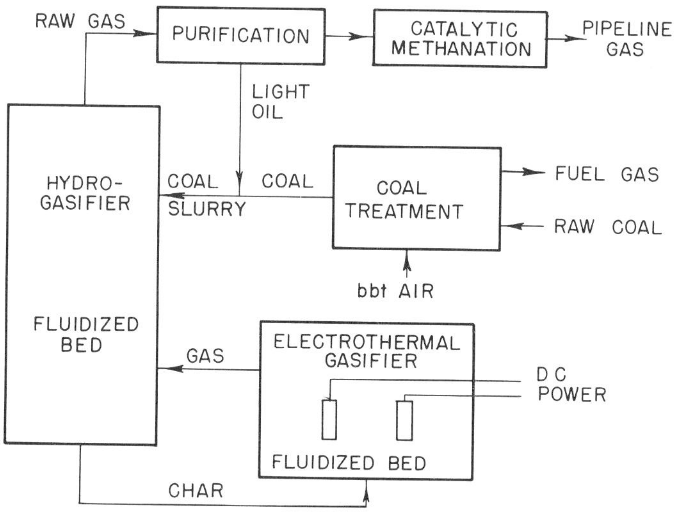

Fig. H-7 Hygas process (*Hottel and Howard, 1971*).

HYPOGENE See Epigene.

HYSTERESIS The lagging of an effect from the expected value due to resistance to movement, such as friction in an instrument or controller, resistance to flow of moisture in a product, or trace of an electrical response in circuits containing magnetic elements.

I

IAEA International Atomic Energy Agency.

IBG Intermediate Btu gas.

ICC Interstate Commerce Commission.

ICGG Illinois Coal Gasification Group (COGAS group).

ICI Imperial Chemical Industries.

ICONS Information Center on Nuclear Standards.

IEA International Energy Agency.

IEEE Institute of Electrical and Electronics Engineers.

IGCC Integrated gasification combined cycle.

IGNITION The process of causing a substance to burn continuously. A phenomenon occurring when the rate of gain of heat of a substance due to oxidation is greater than the rate of heat loss. The temperature of ignition is dependent on the properties of the product and the environment, Table I-1. Some refer to the ignition point as the lowest temperature at which a product will first burn. The flash temperature usually refers to a gas. Also see *Combustion; Flash Point*.

IGNITION POINT See Ignition.

IGNITION TEMPERATURE See Ignition.

IGT Institute of Gas Technology.

IGUA Industrial Gas Users Association.

TABLE I-1
Ignition Temperatures in Air at 1 Atmosphere, Approximate

	Temperature	
	°F	°C
Acetylene	580- 825	305-441
Gasoline (regular-high test)	500- 800	260-427
Hydrogen	1065-1095	574-590
Kerosene	490- 560	255-294
Methane	1170-1380	630-750
Propane	900-1000	480-540
Wood	Approx. 525	Approx. 275
Ethylene	1000-1015	540-545

IIASA International Institute of Applied Systems Analysis.

ILLUMINATING GAS A gas that burns with a luminous flame. Coal gas and water gases are used for illumination while producer gas is used for fuel.

IMPOUNDING AREA An area enclosed by earth banks to contain or impede the flow of water, wastes, tailings, oil, or other liquids.

IMPERIAL GALLON A volume measure used for liquid fuel products in many parts of the world, particularly Great Britain and commonwealth countries, equivalent to 4.55 liters. Also see *Gallon*.

IMPULSE TURBINE See Turbine.

INCIDENT SUNLIGHT The direct sunlight which lands on a surface. The angle of incidence is the angle between the ray of sunlight and a line perpendicular to the surface.

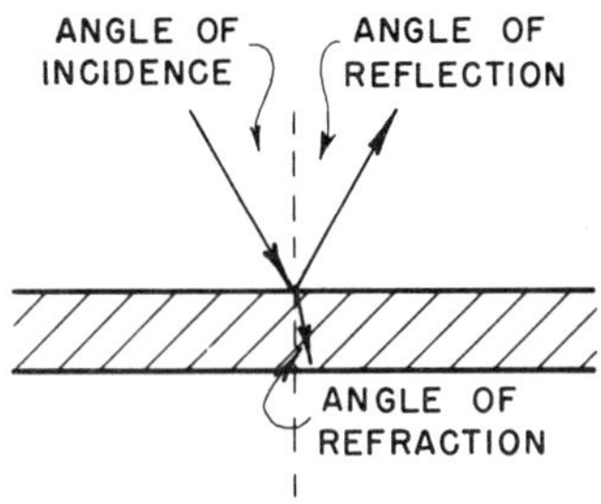

Fig. I-1 Incidence angle.

INCOMBUSTIBLE A substance which will not support combustion or burn at normal temperatures or at specified temperatures.

INDEPENDENT PETROLEUM ASSOCIATION OF AMERICA (IPAA) 1101 16th Street, N.W., Washington, DC 20036.

INDIRECT-CYCLE REACTOR SYSTEM A reactor system in which the primary cooling loop does not include the turbine. In an indirect cycle system, such as a pressurized water reactor, the heat removed from the reactor core in the primary loop is transferred to a secondary loop through a heat exchanger. The secondary loop contains the thermo-mechanical conversion equipment, i.e., the turbine. Indirect cycles have been used in nuclear reactor designs to avoid the reactivity problems introduced by allowing the coolant to boil in the reactor core.

INDIRECT MODE In coal liquefaction, heat is supplied by hot gases from an outside source thus eliminating combustion in the retort.

INDIRECTLY IONIZING RADIATION Radiation consisting of uncharged particles such as neutrons and photons. The uncharged particles traveling through matter do not produce ionization directly but transfer energy to charged particles in the matter which then produce ionization effects.

INDUCED RADIOACTIVITY See Neutron Activation.

INFCE International Nuclear Fuel Cycle Evaluation.

INITIAL CONVERSION RATIO The number of fuel atoms produced from fertile atoms divided by the number of fuel atoms consumed in a nuclear reactor during the early stages of operation of a freshly loaded core. As the fuel in the core is consumed, the conversion ratio changes so that the initial conversion ratio cannot be used directly to determine the final amount of fuel in a used fuel element.

IN-SITU COMBUSTION Burning a material in its location of residence, such as burning coal underground to produce gas which can be used as a fuel. Oil in the oil reservoir can be burned to increase the pressure in the reservoir and to increase the amount of oil recovered. Kerogen can be released from oil shale by *in-situ* combustion.

INCINERATION The burning of materials, usually for the purpose of disposal.

INCOALATION Another word for coalation. See Coalation.

INDICATED HORSEPOWER The power or horsepower of the combustion chamber of an engine obtained by use of a pressure-volume indicator diagram. The numerical value of the indicated horsepower is equal to the friction plus the brake horsepower.

INDIGENOUS   Refers to a material which is native to the geographic area under consideration. Ores and minerals located where they were produced are indigenous in contrast to those which might have been carried to their location.

INDIRECT-FIRED   Those ovens, heaters, dryers, kilns, and devices in which the product being heated receives the heat as a result of transfer through a surface and in which the products of combustion do not reach the product. Also see *Direct Fired*.

INERT GAS   Those gases which do not form compounds under ordinary conditions. Elements which are inert gases are Argon (A), helium (He), Krypton (Kr), neon (Ne), Radon (Rn), and Xenon (Xe).

INDUCED RADIOACTIVITY   The radioactivity developed by a stable nuclide resulting from "bombardment" with artificially produced sub-atomic particles such as neutrons. High energy charged particles produced by an accelerator can also be used to induce radioactivity. Also see *Neutron Activation; Radioactivity*.

INFLAMMABLE   A material which is combustible or has the ability to support combustion, more popularly called flammable. In contrast the word incombustible means lack of ability to burn. Also see *Combustible; Flammable; Limit of Inflammability*.

INFRARED HEATING   Heating through the use of energy within the infrared band of the electromagnetic spectrum, the wavelength band between 0.76 μm to 400 μm. The peak emission at 190°C (374°F) is 6.1 μm; 880°C (1616°F) (gas), 2.5 μm; and at 2200°C (3992°F), 1.1 μm (electric). Infrared energy is given off by all hot objects including electric and gas heaters. Transmission of the energy from the hot surface to the receiving surface depends on characteristics of the surfaces and on absorption characteristics of the gases between the two, particularly water vapor and carbon dioxide. One μm is micrometer, 10^{-4} cm. Also known as radiant heating. Also see *Electromagnetic Spectrum; Infrared Radiation*.

INFRARED RADIATION (IR)   Radiation in the electromagnetic spectrum for which the wavelengths are longer than the visible spectrum and shorter than microwaves, 0.8μ to approximately 1000μ (Figure I-2). Infrared radiation can be subdivided into near IR, intermediate IR, and far IR.

Near infrared	0.8 - 2.5μ
Intermediate infrared	2.5 - 50μ
Far infrared	50 - 1000μ

Also see *Electromagnetic Spectrum*.

INJECTED GAS   The insertion of gas into an oil or gas reservoir formation under pressure, sometimes as high as 140-210 kg/cm^2 (2000

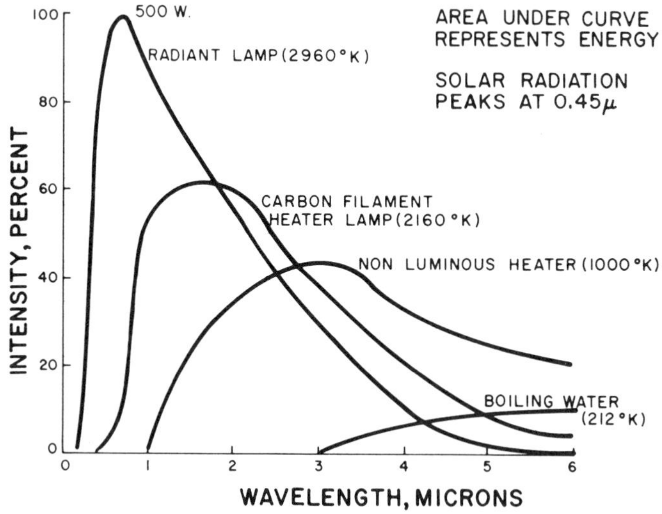

Fig. I-2 Infrared radiation energy from various sources.

to 3000 psi) and often through special wells.

Gas, oil, and water may also be injected into formations as a means of storing them for later use. For example, gas which would otherwise be wasted may be injected into a reservoir for storage.

INORGANIC Material which is not derived from living organisms, plants, animals, or humans. Generally consists of material which does not contain carbon.

INPUT The quantity of material, energy, resource, or power directed into a system either as a batch or as a continuous stream. The input is one of the parameters in determining efficiency. Also see *Efficiency*.

INSOLATION The solar radiation which reaches the earth or a body on the earth, including both direct and sky radiation. The average rate of solar insolation in the United States averaged throughout the year, 24 hr/day, is about 60 Btu/ft^{2}hr or 18 W/ft^2.

INSULATION Material used to isolate or limit the effect or flow of an external phenomenon, such as vibration, sound, heat, voltage (electric), and moisture (vapor barrier).

INSULATOR A material which is a poor conductor of heat and electricity. Generally materials which are good electrical insulators are good thermal insulators. In the atoms comprising a good electrical insulator the outer electrons are tightly bound and are not easily moved.

INTERMEDIATE-BTU GAS See Medium-Btu Gas.

INTERMEDIATE NEUTRON A neutron having kinetic energy above thermal energy (~.025 electron volts) but below the fast range, i.e., less than about 100 kiloelectron volts. Neutrons in the intermediate energy range undergo resonance interactions with many nuclei.

INTENSITY OF RADIATION A measure of the rate of heat transfer by radiation, in units of W/cm^2, J/m^2s, Btu/hr ft^2, etc.

INTERNAL ANGLE OF FRICTION See Friction.

INTERNAL COMBUSTION ENGINE (ICE) An engine in which the fuel is burned in the cylinder in which the piston moves to provide power. Also see *Diesel; External Combustion; Otto Cycle.*

INTEGRATED GASIFICATION COMBINED CYCLE (IGCC) A power plant for production of electricity in which coal is converted to low-Btu gas, which is then used in a combined cycle system to produce electricity (Figure I-3).

INTERNAL ENERGY The energy which a body possesses by virtue of its internal state, represented by U. Internal energy is associated with the translocational energy of molecules and with their individual and collective energy values. U is represented by thermodynamic rather than mechanical parameters. For an ideal gas, $\Delta U = m\ c_v\ \Delta T$. Also see *Energy; Enthalpy; Free Energy.*

INTRINSIC ENERGY Same as Internal Energy.

IR See Infrared.

INTERNATIONAL ATOMIC ENERGY AGENCY (IAEA) An organization of the United Nations, established in 1956, with headquarters located in Vienna, Austria. The objective of the agency is to increase the contribution of atomic energy to health and welfare throughout the world, while providing appropriate control to assure that the energy is not used for military purposes.

INTERNATIONAL SOLAR ENERGY SOCIETY (ISES) An organization of individual members which encourages the development of science and technology, dissemination of information, and use of solar energy. The American headquarters is at 300 State Road 401, Cape Canaveral, FL 32420.

INTERNATIONAL ORGANIZATION FOR STANDARDIZATION (ISO) A nongovernmental organization, which has consultive status with the United

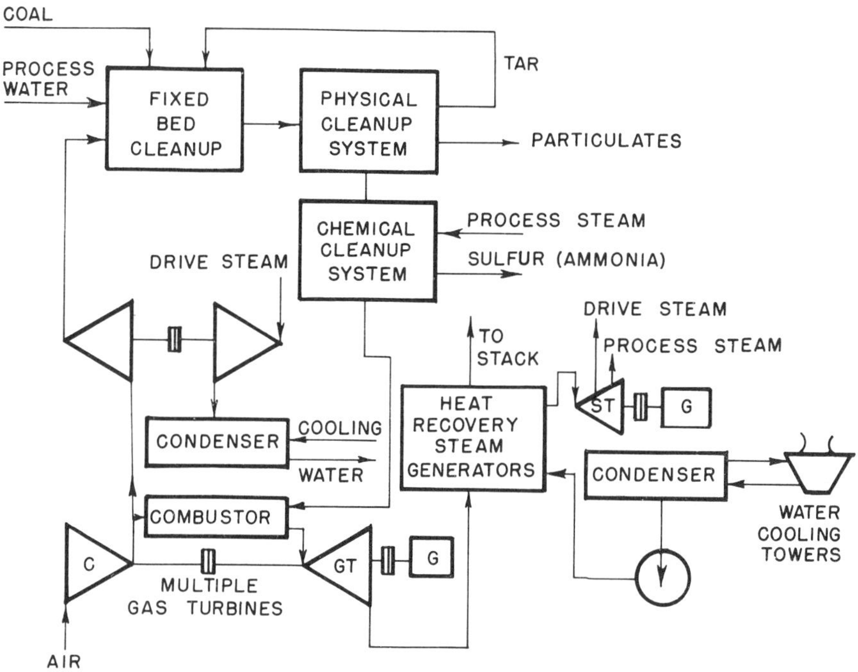

Fig. I-3 Integrated gasification combined cycle (IGCC). *(Corman, 1979.)*

Nations, to "work to develop common industrial, commercial, engineering and safety standards" on a world-wide basis. Located at 1, rue de Varembe, 1211 Geneva 20, Switzerland. Also see *ANSI*.

INVERSE ENERGY EFFICIENCY The total primary energy required to deliver one unit of energy of various types, such as 1.55 units of coal (primary energy) to deliver one unit of liquid. Likewise, 1.55 represents an energy efficiency of conversion of 64.5%.

INVERTER A device or system which can be used to change direct current to alternating current electricity, or vice versa.

ION An atom which has an electric charge produced as a result of the loss or gain of an electron. A positive ion results if an atom loses an electron; a negative ion is formed if an atom gains an electron.

ION ENGINE An engine which propels a spacecraft by emitting energetic ions from the rear of the craft.

IONIC BOND A transfer of electrons between atoms forming ions, with the elements held together by the ionic force or electrovalent bonds. In contrast to the covalent bonds, in which electrons are shared. Also see *Covalent Bond*.

IONIZATION The process of adding or removing electrons from an atom or molecule so that it becomes negatively or positively charged.

IONIZING RADIATION A stream of subatomic particles, e.g., beta or gamma rays, which can produce ions when traveling through matter. Ionizing radiation produces chemical and, in the case of living matter, biological changes in the material through which it passes. If vital organs of the organism are ionized the capacity to grow, reproduce or live is affected. Different levels of exposure produce different results (Table I-2).

TABLE I-2
Approximate Effect of Different Energy Levels of Ionizing Radiation

Product	Energy, rad	Effect
Enzymes	10^7	destroy
Virus	$2\text{-}3 \times 10^6$	destroy
Spore-forming bacteria	10^6	sterilize
Non-spore-forming bacteria	10^5	pasteurize
Insects	$2\text{-}5 \times 10^4$	kill
Insects	10^4	sterilize
Plants	10^3	induce mutations
Man	5×10^2	kill

Rad is a minute quantity of energy of 100 ergs per gram of irradiated material.

Also see *Irradiation; Radiation*.

IPAA Independent Petroleum Association of America.

IRRADIATED FUEL Fuel that has been located in an operating nuclear reactor so that it has been irradiated with neutrons. Fissions produced with fuel by neutron bombardment yield radioactive fission products which make irradiated fuel highly radioactive.

IRRADIATION The process of subjecting a material to some type of radiation such as gamma rays or X-rays. Also see *Gamma Ray; Radiation; X-ray*.

ISABELLE A proton-colliding beam particle accelerator being constructed at Brookhaven expected to be completed in the late 1980's. ISA represents Intersecting Storage Accelerator. Also see *Particle Accelerator*.

ISENTROPIC Pertaining to a thermodynamic process during which the entropy remains constant.

ISES International Solar Energy Society.

ISO International Organization for Standardization.

Iso A prefix meaning equal or similar. For example, isomeric hydrocarbon compounds have the same chemical formula but different arrangements of carbon and hydrogen atoms. Different properties are exhibited by isomers. Also see several words which follow with the prefix iso-.

ISOBUTANE An isomer of butane with the same formula as butane, C_4H_{10}, but with different structure and therefore different characteristics from butane. n-Butane and isobutane are two isomers of butane. The proportion of isobutane produced from catalytic cracking of petroleum is greater than in thermal cracking. n-Butane (normal butane) has a higher boiling point (-0.50°C) than isobutane (-11.7°C) at atmocpheric pressure.

	Butane	Isobutane
Boiling point, °C (°F)	0.5 (32.9)	-11.7 (10.9)
Solidification point, °C (°F)	-138 (-216)	-160 (-256)
Density, g/ml	0.579	0.557

ISOMER One of two or more compounds which have identical composition and formula but different structure. The physical properties of different isomers are different. Examples of hydrocarbon isomers derived from petroleum are: Butane, isobutane: (C_4H_{10}); pentane, isopentane: (C_5H_{12}); octane, isooctane: (C_8H_{18}); butylene, isobutylene: (C_4H_8). Also see *Octane*.

ISOOCTANE An isomer of octane, C_8H_{18}, structured in the shape of a spider web. It is a colorless liquid, with a boiling point of 99°C (210°F), which is manufactured by reacting gas from a catalytic cracker with gas from a catalytic reformer in the presence of sulfuric or hydrofluoric acid. Isooctane is added to gasoline to decrease knocking tendencies during combustion in an internal combustion engine. Also see *Isomer; Octane; Octane Number*.

ISOPARAFFIN SERIES One of the principal series of hydrocarbons, a branched-chain paraffin. Also see *Hydrocarbon Series*.

ISOPIESTE A curve or line representing equal moisture contents, or other appropriate parameters, plotted for a constant pressure.

ISOSTERE An equation or curve representing moisture content relationships plotted for a constant moisture content.

ISOTONIC The exertion of the same osmotic pull by one medium as the other medium on the opposite side of a semi-permeable membrane. Thus, the state where the same concentration of particles exists on both sides of the semi-permeable membrane between two fluids.

ISOTHERMAL A constant temperature process. Isothermal lines connect points at the same temperature.

ISOTOPE One of two or more nuclides of an element having the same number of protons in their nuclei. The chemical elements and their isotopes are identified by atomic number (Z). Isotopes have different mass numbers (A) because of differences in the numbers of neutrons.

Isotopes of Carbon

	Z	A
$_6C^{12}$	6	12
$_6C^{13}$	6	13
$_6C^{14}$	6	14

Isotopes may be stable, and thus not radioactive, or may be radioactive, known as radioisotopes. Also see *Atomic Number; Nuclide; Radioisotope; Radionuclide.*

ISOTROPIC ENRICHMENT A process which alters a substance to increase the concentrations of one isotope compared to another. The uranium used in light water reactors is isotopically enriched in ^{235}U.

ISOTROPIC The condition in which the same properties exist in a material in all directions from a point or location.

IUPAC International Union of Pure and Applied Chemistry.

i-x DIAGRAM A form of the Mollier diagram in which the axes are the enthalpy and the absolute humidity of the air with other lines or curves representing the pressure and temperature. The i-x axes are not necessarily at right angles.

J

JACK PLANT A central plant for pumping oil which provides power and connections with cables or rods going to each well to operate several pumps.

JAY A term used for inferior cannel coal or kennel coal. Also see *Cannel Coal; Kennel Coal.*

JCAE Joint Committee on Atomic Energy.

JET A black lignite which will take a good polish.

JET ENGINE An engine for providing jet propulsion, the principal categories being pulse-jet, ram jet, turbojet, and turboprop for airplanes. Jet engines can use a wide variety of fuels.

JET FUELS Fuels used for aircraft turbine engines and turbines for power plants. Jet fuel grades are designated by JP followed by a number such as JP-1, JP-2, JP-3, JP-4, JP-5, JP-6 or by jet followed by a letter, such as Jet A or Jet B. Also see *Reid Vapor Pressure.*

JOFFE BARS A special type of magnetic field placed around fusion plasma in an open system to keep it stable.

JOJOBA A desert shrub which offers potential as an energy source.

JOULE The amount of work expended when a force of one newton (10^5 dynes) acts through a distance of one meter. One joule equals 10^7 ergs or 0.7375 ft lb or 0.7390 gram calories. Also see *Erg; Work.*

JOULE CYCLE See Brayton Cycle.

JOULE LAW An equation which gives the heat generated in joules, H, when a current passes through a conductor:

$$H = I^2Rt$$

where

I = electric current, amperes
R = resistance, ohms
t = time, sec.

JOULE-THOMSON EFFECT The cooling which results when a highly compressed gas is permitted to expand adiabatically such that no work is done.

JUDD A block of coal ready for fracturing or grinding to reduce size.

K

k   A letter commonly used to represent the thermal conductivity of a material in heat transfer calculations. The constant for an ideal gas or mixture of gases, sometimes called adiabatic exponent, also represented by γ.

K, KELVIN   An abbreviation and scale of absolute temperature. 273 K is equivalent to 0°C and 32°F. Absolute zero is -273°C and 373 K is equivalent to 100°C and 212°F. In SI nomenclature, K represents kelvin temperature without the degree symbol. Also see *Absolute Temperature; SI Units.*

KAON   A type of meson. There are positive, negative, and zero charged kaons, which have masses of approximately 500 Mev. The kaons are not stable. K^+ and K^- have mean lives of approximately 1.2×10^{-8} sec. The two K° mesons, K_S^6 and K_L^6 have mean lives of about 0.86×10^{-10} sec.

KAPLAN TURBINE   Consisting of a propeller with adjustable blades. A reaction hydraulic turbine. It is used at hydroelectric projects with low heads, 15 m (50 ft). Also see *Dariaz Turbine; Francis Turbine; Hydraulic Turbine; Pelton Wheel.*

KATA-   A prefix, which is also used as cata, used to indicate the deepest zone of metamorphism, in contrast to epi-, the prefix for upper layer, and meso-, the prefix for middle layer.

KELLOGG PROCESS   A process for gasification of coal in an oxygen-steam environment under pressure in the presence of a catalyst, sodium carbonate, to produce synthetic gas. Also see *Hydrane Gasification; Koppers-Totzek Process.*

KELP   A brown algae plant of the *Laminariales* order with large blade-shaped or vine-like structure which grows in bodies of water and is a potential source of biomass for energy (Table K-1). The species most often referred to are elk, bull, and giant kelp.

TABLE K-1
Kelp Genus, Location and Size

	Genus	Location	Size
Elk	*Pelagophycus*	California + Baha California	up to 100 ft long
Bull	*Nereocystis*	S. California to Alaska	100 ft and longer
Giant	*Macrocystis*	Oceans of the world	up to 150 ft long

Also see *Brown Algae; Bull Kelp; Giant Kelp.*

KELVIN TEMPERATURE SCALE An absolute scale of temperature, K, which is related to the Celsius scale. The absolute temperature in (K) is the temperature in degree Celsius (°C) plus 273.15. Also see *K; Temperature Scale.*

KENNEL COAL A coal that can be ignited at the temperature of a match flame and will burn with a bright flame. Also called candle coal or cannel coal.

KERMA The dose delivered to matter by uncharged particles (neutrons, photons, etc.) as recoil kinetic energy of the charged particles with which the uncharged ones interact. The kerma is measured as initial recoil kinetic energy per unit mass of matter.

KEROGEN A bituminous substance which is the precursor to oil in the formation of oil shale, made up of a combination of compounds. It yields an oily product when heated in the absence of oxygen. Kerogen is a high molecular weight polymer, one example from Colorado oil shale, has the following contents:

Carbon	80.5%
Hydrogen	10.3
Oxygen	5.8
Nitrogen	2.4
Other	1.0

Also see *Bituminous; Oil Shale.*

KEROGEN SHALE See Bituminous Shale.

KEROSENE; KEROSINE A liquid used as a fuel which is obtained from petroleum in the fractional distillation of oil after the gasoline has been removed. It is a light oily liquid with a net heating value of approximately 43,270 kJ/kg (18,600 Btu/lb). Kerosine was formerly obtained from destructive distillation of bituminous coal, hence is often called coal oil.

KGRA Known geothermal resource area.

KILO CALORIE See Calorie.

KILOGRAM A unit of weight in the metric system equal to 1000 grams or 2.2046 lb.

KILOGRAM-CALORIE See Calorie.

KILOMETER A unit of measure of distance in the metric system which is 1000 meters or 0.62137 miles.

KILOTON ENERGY A measure of the energy released during a chemical or nuclear explosion. The kiloton unit is the amount of energy released by combustion of one kiloton of TNT. The atomic bombs dropped on Hiroshima and Nagasaki each released approximately 20 kilotons of energy. Also see *Megaton*.

KILOWATT A unit of power equal to one thousand watts or 10^{10} dyne-cm/sec; 746 watts is equal to 1 horsepower.

KILOWATT-HOUR Represented by kwh, kwhr, kwHr, or kWh, a unit of energy consisting of power used over a period of time in hours used in commercial applications; 1000 watt-hr is equal to 1 kilowatt-hour. 1 kWh = 1.340 hp.

KINDLING TEMPERATURE The temperature at which rapid oxidation and burning of a material occurs. Also see *Combustion; Ignition*.

KINETIC ENERGY The energy possessed by a body by virtue of its motion, K.E. = $1/2\ mv^2$, where m is the mass of the body and v the velocity. Also see *Energy*.

KIR A designation given to petroleum possessing the appearance of asphalt, which has solidified when exposed to the atmosphere.

KNOCK The noise resulting from self-ignition in the operation of an internal combustion engine, usually undesirable.

KOPPERS-HASCHE PROCESS A process for reforming natural gases to form gases of lighter constituents by using air to provide partial oxidation in the presence of an alumina catalyst. Also see *Cracking; Reforming*.

KOPPERS-TOTZEK PROCESS A process for gasification of coal in an environment of oxygen-steam at atmospheric pressure. Used for the production of ammonia and methanol. The gas produced can be upgraded to pipeline quality by methanation (Figure K-1). Also see *Hydrane Gasification; Kellogg Process*.

KRYPTON 85 A radioisotope of krypton produced in nuclear reactors. ^{85}Kr has a half-life of 10.7 years. Krypton, a noble gas, is a

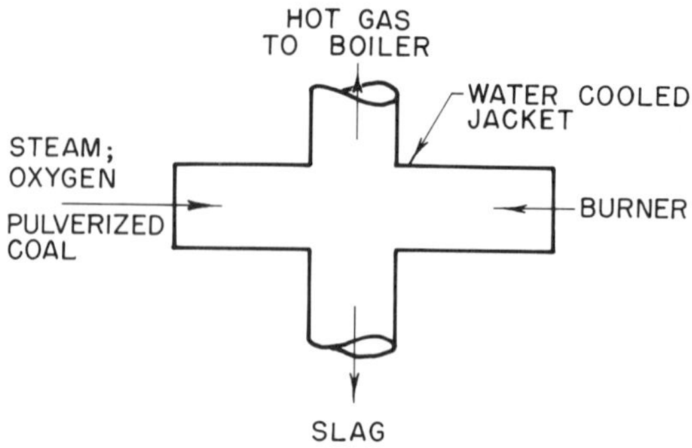

Fig. K-1 Koppers-Totzek gasifier.

radioactive waste product of reactor operation which poses special disposal problems because of its chemical inertness.

kVDC Kilovolts, direct current.

kWh, kwh, KwHr, KWH See Kilowatt-Hour.

L

L, ℓ Liter

LABELING Incorporating radioactive nuclides into some of the molecules of a compound so that the compound can be traced through a series of biological and chemical reactions. Common labels in biological research are the hydrogen isotope called tritium, ^{3}H, and radiocarbon, ^{14}C. Also see *Tritium*.

LABILE Used to refer to constituents in rocks, minerals, and plant and animal products which decompose easily, especially in a normal environment, or change readily by heat or oxidation. Also called metastable.

LAGGING Applies to many energy-related activities, such as the insulation on pipes, boilers, and heat transfer equipment; the securing of walls and roofs in mines and tunnels; and materials placed on surfaces to increase friction such as on drums and pulleys.

LAGOON A shallow body into which liquid waste is placed with the objective of having biological activity reduce the organic content. The effluent has a low biological content. Lagoons may be placed in series, may incorporate an aeration device, and may be treated to assist the biological processes and the removal of solids.

LASER A device that produces a coherent beam of monochromatic light. Powerful lasers can deliver intense bursts of energy to small volumes. In fusion research, they are being used to attempt to raise the temperature of small pellets of deuterium-bearing materials to the point where thermonuclear reactions will take place.

LASER INDUCED FUSION Fusion of nuclei that is initiated by compressing the target material (fusion fuel) to very high density using focused, pulsed laser beams. The laser beams ablate the surface of the target pellet, typically a tiny hollow glass sphere containing tritium and deuterium gases. The reaction impulse to the outward momentum of the ablating material compresses the remaining unablated pellet material. A new target pellet is required for each pulse of energy produced. This concept does not require magnetic confinement of the high temperature plasma in order to produce fusion energy.

LASL Los Alamos Scientific Laboratory.

LATENT HEAT The quantity of heat released or absorbed during a phase change. For example, the quantity of heat required to change a substance from a solid to a liquid at constant temperature is the latent heat of fusion (Table L-1). To change a substance from a liquid to a vapor at constant temperature requires the addition of heat, called the latent heat of vaporization. The reverse process is accomplished by the release of the same quantity of heat.

TABLE L-1
Latent Heat of Water

For water	Metric units	English units	SI units
Latent heat of fusion	79.1 cal/g (at 0°C)	144 Btu/lb (at 32°F)	334.9 kJ/kg (at 0°C)
Latent heat of vaporization (or evaporation)	539.55 cal/g (at 100°C)	970.4 Btu/lb (at 212°F)	2257.0 kJ/kg (at 100°C)

LATENT HEAT STORAGE A method of storing heat either as latent heat of fusion (liquid-solid) or latent heat of vaporization (gas-liquid). Latent heat absorption per unit weight is usually much greater than sensible heat absorption over a reasonable temperature range. Glauber's salt (Na_2SO_4 + H_2O) has been a medium for heat storage in the normal temperature range.

LATTICE A regular array. The term is often used to describe the array of atoms in a crystalline material or the array of fuel elements or fuel clusters in the core of a nuclear reactor.

LAVA See Magma.

LBG Low-Btu gas.

LCRM Low cost residual material.

LEAD A dense soft metal represented by the chemical symbol Pb. Lead is used as a shield for ionizing radiation, a constituent in the engine knock suppressants tetraethyl lead (TEL) or tetramethyl lead (TML), and for lead-acid batteries.

The properties of lead are: Atomic number, 82; atomic weight,

207.19; density, 11.350 kg/m^3 (11.35 g/cm^3); melting point, 327.5°C (622°F); and boiling point, 1740°C (3164°F). Also see *Lead-Acid Battery; MEL; MTPB; Octane Number; TML.*

LEAD-ACID BATTERY A battery in which electric energy is stored. This type of battery commonly is used in 6- and 12-volt automotive systems for starting, controlling, and lighting. The negative plates and terminal are lead; the positive plates are coated with lead dioxide. The plates, with sulfuric acid between, form a cell.

LEADED GASOLINE Gasoline to which a lead compound is added to permit a higher compression ratio for combustion without knocking. Also see *Tetraethyl Lead; Tetramethyl Lead.*

LEAKAGE The flow of neutrons out of the core of a nuclear reactor. Some of the neutrons produced by nuclear fission in a reactor core migrate out of the core before they are absorbed. These neutrons are said to leak out of the core. The migration process involves a sequence of straight line trajectories, each terminated when the neutron scatters from a nucleus and begins a new trajectory. Some of the neutrons that leak out of a reactor core return to it following scattering reactions that redirect the neutrons back in the core. In fact, a reactor often has a structure just outside the core, called the reflector, whose purpose is to reflect leakage neutrons back into the core.

LEAN GAS Used to describe gas with heat content of 500 Btu/ft^3 or less. Also see *Low-Btu Gas; Medium-Btu Gas.*

LEL Lower explosive limit.

LEPTON Any of a number of fundamental particles that form a family with certain similar properties. In general the leptons are light particles that do not interact strongly with the members of the other principal family, the hadrons. The common members of the lepton family are electrons, neutrinos, mu mesons, and their antiparticles.

LET Linear energy transfer.

LETHAL DOSE A dose of radiation that will cause death. The response of organisms to radiation varies widely among species and indeed among individuals of given species so that no single number can be given for a lethal dose. In humans a dose of several hundred rem received by the whole body will cause death within days to a substantial fraction of those who receive it.

LGR Light-water cooled, graphite-moderated reactor.

LHV Low heating value.

LIFE, LIFETIME The period of time during which an initial amount of radioactive material diminishes, by decay, to a fraction of the initial amount. If the fraction is one-half, the lifetime is the half-life. If the fraction is e^{-1}, the lifetime is the mean life. Both lifetime measures are commonly used.

The lifetime concept is also used in a probability interpretation relating to the decay of an individual unstable particle or nucleus. The half-life in this case is the time required to reduce to one-half the probability that an unstable particle or nucleus, remains undecayed. A similar definition applies to mean life. Also see *Half-Life.*

LIFE CYCLE COSTING A procedure which is a part of the design process in which the costs of various systems are compared throughout the estimated life of the product. Life cycle costing has become more significant as the cost of operation of a product or system has increased, primarily as a result of higher fuel costs.

LIGHT OIL Oil distilled from coal tar at temperatures below 200°C (392°F). Also see *Heavy Oil.*

LIGHT WATER Water which is composed of oxygen and the light isotope of hydrogen ($^1H^1$). Water composed of oxygen and deuterium is called heavy water.

LIGHT WATER REACTORS (LWR) The principal types of nuclear reactors in the U.S.A., BWR and PWR use light water as the heat transfer medium (Figure L-1). Also see *Boiling Water Reactor; Light Water; Nuclear Fuel Cycle; Pressurized Water Reactor.*

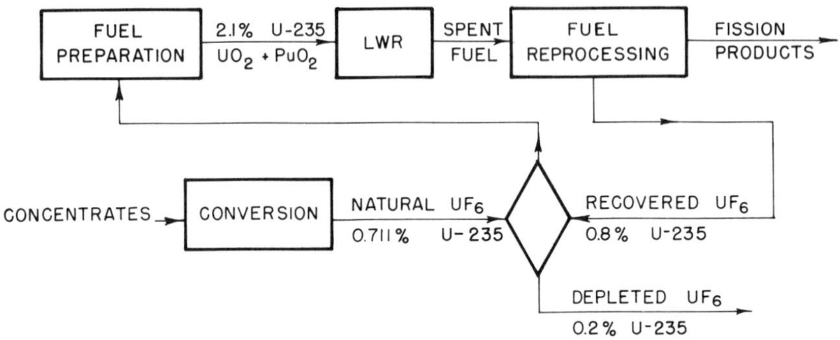

Fig. L-1 Light water reactor cycle *(Duderstadt, 1979).*

LIGNIN A non-carbohydrate polymeric material in the fibrous portion of wood and vascular plants which gives structure to the plant and supports the cellulose, which has the general formula $C_9H_{10}O_3(OCH_3)_n$, n equal 0.9-1.7. Some references give a formula of

$C_{40-41}H_{40-44}O_{12-16}$. Lignin has a molecular weight of 2000 to 15,000. Principal sources of biomass are the lignocellulose materials. Also see *Biomass; Wood.*

LIGNITE A brown-black coal in which the fossilization of organic materials has proceeded further than with peat but not as far as with sub-bituminous coal. Lignite contains about 30 to 40% water and has a heat value of about 13,900 to 19,290 kJ/kg (6000 to 8300 Btu/lb) on a maf basis. The quantity of known lignite reserves is about one-fourth of the coal reserves in the United States. Lignite is used for operating power plants and for industrial heating. It has a great potential for gasification to produce quality gas. Also see *Lignitous Coal.*

LIGNITE SHALE See Bituminous Shale; Lignitous Coal; Lurgi Process; Mature Lignite.

LIGNITOUS COAL Coal which has a carbon content of 75% to 84% on a moisture ash-free basis (maf). In contrast lignitic coal, lignitous coal, is classified as sub-bituminous coal.

LIGNOCELLULOSE A group of substances in woody plants consisting primarily of lignin and cellulose. Principal lignocellulose materials for biomass are trees, woody plants, straw, forage, and other plant residues. These materials can be burned or converted to gases or liquids for fuel. Also see *Cellulose; Hemicellulose; Holocellulose; Lignin.*

LIMIT OF INFLAMMABILITY The high temperature below which and the low temperature above which (borderline values) a mixture of combustible gas and air sustain combustion for a particular fuel or product. Also see *Explosion, Gases; Flammable; Inflammable.*

LIMIT SWITCH A device which operates to limit a function by turning off or on an operation. The limit switch can operate to control a lower or upper limit. Limit switches are commonly used on burner controls for furnaces and boilers, heating and cooling equipment, volume devices such as cylinders and tanks, pressure devices to limit maximum pressure as well as to assure a minimum pressure, such as pressure of gas or oil supply to a burner.

LINEAR ENERGY TRANSFER (LET) The transfer of energy along the path of an ionizing particle. Energy is transferred from the particle to the medium through which it is moving.

LIQUATION A process of separating materials in which one material has a different temperature of melting than another or other materials. The process is used to recover sulfur.

LIQUEFACTION (or LIQUIFACTION) The conversion of coal or other solid into a liquid hydrocarbon and related compounds by hydrogenation;

the conversion of a gaseous material into a liquid; or the changing of organic matter in waste to a soluble state. Also see *Coal Liquefaction; H-Coal.*

LIQUEFIED NATURAL GAS (LNG) Natural gas which is liquefied by compression and cooling requiring about 15% of the energy of the gas. Liquefied natural gas is being transported across oceans in tankers and supertankers. The entire process will deliver approximately 70% of the energy to the consumer. Also see *Natural Gas.*

LIQUEFIED PETROLEUM GAS (LPG) A mixture of gaseous petroleum products normally stored and transported as a liquid under pressure. The principal constituent is propane, but may also include butane and pentane. Liquid petroleum gases are usually taken from natural gas or may be obtained from distillation of crude oil. LP gas has an ignition temperature of approximately 450°C (810°F) and a net heating value of 45,300 kJ/kg (19,500 Btu/lb). Also see *Butane; Liquefied Natural Gas; Propane.*

LIQUID METAL (for nuclear applications) Sodium, potassium, or a combination of the two called NaK that are considered to be possible nuclear reactor coolants. Neither sodium (m.p. 98°C) nor potassium (m.p. 64°C) is liquid at room temperature but have melting points that are low enough to make them useful as reactor coolants.

LIQUID METAL-COOLED FAST BREEDER REACTOR (LMFBR) A fast breeder nuclear reactor which uses liquid metal, such as sodium for a coolant which becomes radioactive. Enriched uranium-235 or plutonium-239 is used as a fuel (Figure L-2). A breeder reactor produces fuel as it consumes fuel. A separate circuit using sodium is used to transfer heat from the core to the process, usually for steam generation. Also see *Breeder Reactor; Fast Breeder Reactor.*

LIQUID OXYGEN (LO_2, LOX) Oxygen which is in a liquid state obtained by compressing oxygen or by maintaining oxygen under pressure used to support combustion or respiration. Liquid oxygen is used as a rocket fuel to supply the oxygen for combustion in space. Ethyl alcohol, liquid hydrocarbon fuels (kerosene), and ammonia fuels in rockets normally use liquid oxygen as oxidizers. Liquid oxygen has a low boiling point of 130°C (-297°F), a density of 1.142 g/ml at the b.p., and a critical pressure of 50 atm. Also see *Oxygen.*

LITER (or LITRE) A unit of volumetric measure in the metric system of one kilogram of water at 4°C (39.2°F) and 760 mm pressure. One liter has a volume of 0.001 m^3, 1.000028 cubic decimeter (dm^3), 61.026 in^3 or 1.0567 liquid quarts or 0.26 gal. Also see *Conversion Units; Gallons.*

LITHIUM A light metal, Li, having a density approximately half of water. When irradiated with neutrons, lithium yields tritium, a product placed in the atmosphere from explosion of the hydrogen bomb,

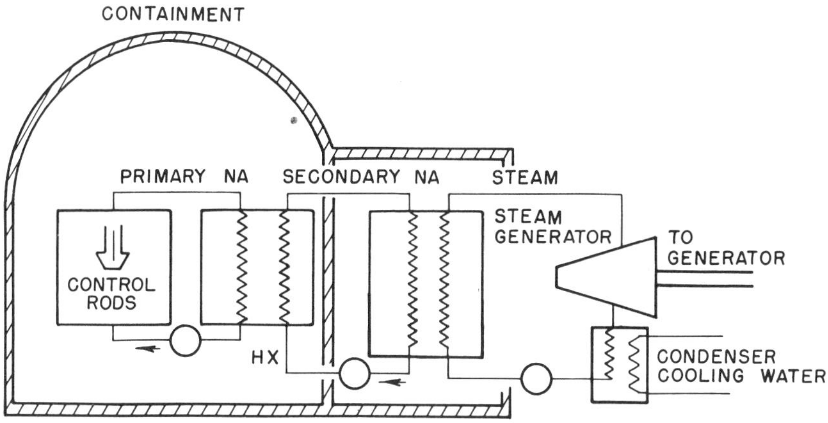

Fig. L-2 Liquid metal fast breeder cycle (LMFBR).

a fusion process. With the testing of hydrogen bombs in the atmosphere in the 1950's, the presence of tritium as in underground water, shows that water from the surface has entered the underground storage. The properties of lithium are: Atomic number, 3; atomic weight, 6.9; density 534 kg/m^3 (33 lb/ft^3); melting point, 180°C (356°F); and boiling point, 1347°C (2457°F). Also see *Fusion; Tritium.*

LITHOSPHERE See Earth; Geosphere; Mantle.

LIVE OIL Crude oil that contains natural gas.

LIVE STEAM The name given to steam as it leaves the boiler before heat has been removed. After heat has been removed by the operation it is called exhaust steam, which still contains considerable heat, but which is not always easy or economical to utilize.

LLL Lawrence Livermore Laboratory.

LLRW Low level radioactive waste.

LMFBR Liquid metal fast breeder reactor.

LNG Liquefied natural gas.

LO_2, LOX Liquid oxygen (see).

LOAD FACTOR As applied to an electrical load the ratio of the average to the peak over a given period of time, a value less than 1 or less than 100%.

LOAD LEVELING The procedure of adjusting demand for electricity to reduce peak requirements and thus to provide a more uniform load to be supplied by the utility supplier.

LOCA Loss of coolant accident.

LOFT Loss of Flow Test Facility.

LONGWALL MINING A method for removing and conveying coal from underground (deep) mines. In longwall mining a mechanical device moves approximately 500 ft along the wall of the mine with a minimum of manpower and under safer conditions than are experienced with conventional mining. The process is more economical than conventional mining and leaves less coal in the ground.

LOSS The amount or portion of energy, force, power, pressure, or product lost during an operation or process such as manufacturing, processing, drying, storage, handling, packaging, etc. Usually an increase in loss results in a reduction of efficiency.

LOSS OF COOLANT ACCIDENT (LOCA) A nuclear reactor accident initiated when the coolant flow to the core is interrupted.

LOSS OF FLOW TEST FACILITY (LOFT) An experimental facility, located in Idaho, that is used to test the consequences of a loss of coolant flow in the core of a nuclear reactor.

LOW-Btu GAS A gaseous fuel having a heat value below 13 MJ/m^3 (350 Btu/m^3). The fuel includes carbon monoxide and other gases which are byproducts of industrial processes. Producer gas 5.2 MJ/m^3 (140 Btu/ft^3) and blast furmace gas 3.3 MJ/m^3 (90 Btu/ft^3) are low-Btu gases. Also see *Blast-Furnace Gas; High-Btu Gas; Medium-Btu Gas; Producer Gas.*

LOW-GRADE HEAT Thermal energy at a temperature below 100°C (212°F). Very large quantities of low-grade heat are available as waste heat from industrial processes. Low-grade heat has a low availability factor. Also see *Availability Factor; Geothermal Energy.*

LOW-HEAD HYDRO The use of low-head dams to produce hydroelectric power, defined by DOE as dams of 20 m (66 ft) or less. Approximately 54,600 MW of additional electric generating capacity is available by further developing present dams in the U.S.A. Turbines used are those suitable for low head applications.

LOW-HEATING VALUE The energy output of a fuel, Joules or Btu per unit weight or volume, under conditions where water vapor rather than liquid water is a combustion product (Table L-2). The low heating value is usually more useful than the high heating value for applications because the exhaust temperature of a combustion unit is usually above the vaporization temperature of water so that the heat in the moisture in the products of combustion is not utilized. The low heating value, Btu/lb of fuel, based on the components is

$$\text{Low heating value} = 14{,}400\ C + 52{,}000\ (H - \frac{O}{8}) + 4{,}500\ S + 1{,}100\ w$$

where

C = the weight of carbon/lb of fuel
H = weight of hydrogen/lb of fuel
O = weight of oxygen/lb of fuel
S = weight of sulfur/lb of fuel
w = weight of moisture/lb of fuel

For conversion to SI units, 1 Btu/lb = 2.324 kJ/kg. The low heating value is also known as the net heating value. Also see *High Heating Value.*

TABLE L-2
Low Heating Value of Various Fuels, Approximate

		Low heating	Value
	Density	Btu/lb	kJ/kg
Coal			
Anthracite		up to 14,000	32,480
Bituminous		12,950	29,785
Lignite		6,000	13,920
Fuel Oil	lb/gal		
No. 1	6.79	18,690	43,470
No. 2	7.21	18,310	42,590
No. 4	7.78	17,740	41,260
No. 5	7.83	17,640	41,030
No. 6	8.05	17,180	39,960
Gasoline	6.15	18,910	43,960
Kerosene	6.82	18,660	43,400
Gas (C_nH_{2n+2})	lb/cu ft		
Natural (CH_4) methane	0.046	22,190	50,010
Ethane (C_2H_6)	0.085	20,416	47,480
Manufactured	0.048	10,580	24,610
Propane (C_3H_8)	0.12	19,944	46,390
Butane (C_4H_{10})	0.15	19,680	45,775
Wood, dry		8,000	18,600

LOW-LEVEL RADIATION DOSE RATES The dose rates from exposure to radiation in the U.S.A. (1979) (Table L-3).

LOW-LEVEL RADIOACTIVE WASTES (LLRW) All wastes containing radioactive materials except for (1) high-level radioactive wastes (HLW)

TABLE L-3
Dose Rates from Exposure to Radiation, millirem yr

Maximum allowable to industrial workers	5,000
Intercontinental flight crew (70 hr/month)	385
Nuclear power occupational	365
Average natural background, Colorado	170
Average natural background, U.S.A.	100
Medical diagnostics	70
Cosmic rays, terrestrial sources	35
Washington, D.C., per working year	20
Nuclear weapons fallout	4.4
Natural gas, residential use	2
TMI exposure, 50 mi radius, average	1.5
Nuclear power (1980)	0.1

that include the concentrated fission products from reactors and (2) transuranic wastes that contain transuranic elements extracted from spent fuel elements. Low-level radioactive wastes contain less than 10 nanocuries of transuranic activity per gram and/or may have low concentrations of radionuclides. The present annual production of low-level radioactive waste is about 0.45 kg/person (113,200 m^3). *(Gloyna, 1980.)* Also see *High-Level Radioactive Waste; Transuranic Wastes.*

LOW PRESSURE AIR BURNER A burner which operates at a pressure under 0.34 atm (5 psi).

LOW SULFUR COAL Coal which contains less than two-thirds of a pound of sulfur for each million Btu of heat produced.

LOW SULFUR CRUDE OIL Crude oil or petroleum which contains less than 0.5% of sulfur by weight.

LPG Liquefied petroleum gas or LP-gas.

LRA Land resource area (for biomass).

LRR Land resource region.

LSI Large-scale integration (of electric circuits).

LUBRICANTS Oils used to reduce friction of moving parts. They are usually made by vacuum distillation of the residues from refining petroleum.

LUMEN See Candela.

LURGI PROCESS A method of gasifying coal and lignite using oxygen and steam under pressure of about 35 atmospheres, with process temperatures from 650° to 850°C (1200° to 1560°F). Synthesis gas is produced which can be converted to pipeline gas (Figure L-3).

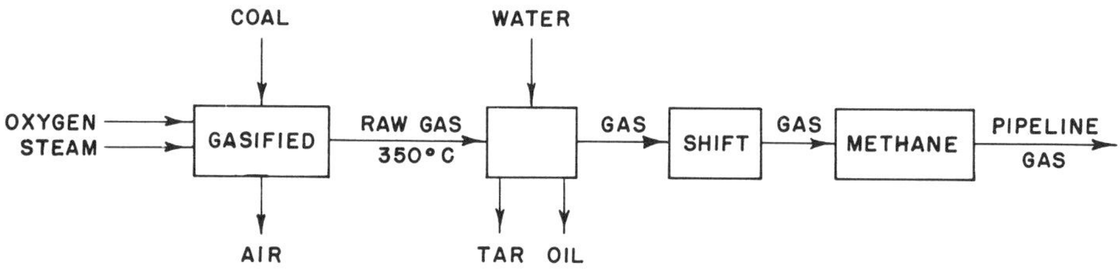

Fig. L-3 Lurgi process of gasification.

Also see *Fischer-Tropsch Process*.

LWR Light water reactor.

M

M   Used to represent mega or million or, which is confusing, thousand for some uses such as Mft^3 natural gas and Mbm board feet.

MAF, maf   Moisture and ash-free.

MAGMA ENERGY   An energy source in the earth with heat in the molten rock, liquids and gases, or magma below the earth's crust. The magma is at a temperature of 982° to 1093°C (1800° to 2000°F), much hotter than the geothermal sources which are at 260°C (500°F) or less. Magma becomes lava when it pours from the surface of the earth at volcanoes and releases its dissolved gases. Also see *Earth; Geothermal; Geothermal Gradient; Hot Rock.*

MACROCLASTIC COAL   A laminated coal which has a high proportion of layers of vitrinite as well as resins and spores.

MAGNETIC BOTTLE   A magnetic field configuration that is effective in confining charged particles. The space to which the particles are confined is called the bottle. Magnetic bottles tend to be leaky so that confinement times are limited.

MAGNETIC CONTAINMENT   Also referred to as a magnetic bottle, a device using a magnetic field for confining plasma to a particular volume and location for controlled fusion in a thermonuclear reactor. Also see *Fusion; Magnetic Bottle; Plasma.*

MAGNETIC PRESSURE   A magnetic force that confines a plasma to a limited volume in the same way that it would be confined if it were a neutral gas under pressure. The equivalent pressure is called the magnetic pressure.

MAGNETIC PUMPING   A process of adding energy to a plasma using time-dependent magnetic fields. A changing magnetic field produces an electric field that can accelerate the charged particles of a plasma.

MAGNETOHYDRODYNAMIC CONVERSION OR POWER GENERATION (MHD) A procedure for converting thermal energy at a high temperature to electricity (DC) by passing an ionized gas (plasma) through a fixed magnetic field. A thermal efficiency of conversion is about 25-30%. This is considered an indirect conversion because a working fluid is involved between the heat application and the production of electricity in contrast to direct conversion by thermionic and thermoelectric methods. MHD may be used to obtain additional power from conventional fossil fuel power plants, and possibly from nuclear power plants. The open cycle system, as contrasted to the closed cycle system, is applicable to fossil fuel power plants. Also see *Thermionic; Thermoelectric.*

MAJOR/MINOR MINE DISASTER A principal criterion for identifying the extent of a mine disaster is the number of people killed. If five or more people are killed it is a major disaster; if fewer, a minor disaster, depending also on the extent of property damage.

MALTING COAL A term used in Wales for anthracite coal. Also see *Anthracite Coal; Hard Coal.*

MANHATTAN PROJECT The code name for the research and development project that developed the atomic bomb during World War II. The principal sites for the project were Chicago, Illinois; Oak Ridge, Tennessee; Hanford, Washington; and Los Alamos, New Mexico.

MANIFOLD A pipe or chamber which is joined to a series of other pipes. Also see *Mixing Chamber.*

MANTLE The region of the earth between the crust and core. The mantle begins about 35 km below the surface on land and 10 km below the ocean (Figure M-1). Little is known directly about the mantle. Also see *Atmosphere; Earth; Heat Flow, Earth; Lithosphere; Mohole.*

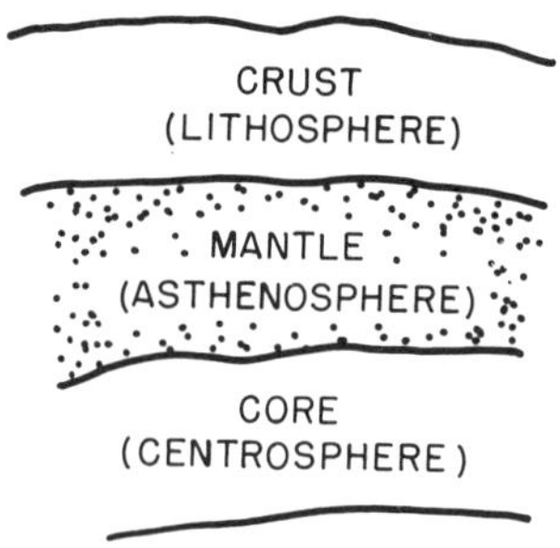

Fig. M-1. Mantle.

MAOP Maximum allowable operating pressure.

MARICULTURE The study and cultivation of marine life; the bodies of water supporting marine life, plants, fish, and animals, are potential sources of organic matter for energy for conversion to food and fuel. Also see *Aquaculture*.

MARSH GAS Gas which emanates from marshes, swamps, and wet lands, as a result of decay of organic matter under anaerobic conditions. The gas is principally methane. Also see *Biogas; Firedamp; Methane*.

MASS ENERGY EQUIVALENCE (Relation) The principle, first proposed by Einstein, that mass and energy are interchangeable. The equivalence is expressed quantitatively by the equation $E = mc^2$, where E is the energy equivalent to mass m and c is the speed of light in vacuum.

MASS NUMBER (A) An integer equal to the total number of protons and neutrons in a nucleus. The mass of a nucleus consists almost entirely of the masses of its neutrons and protons. Because the masses of neutron and proton are almost the same, the mass number as defined above is a good indicator of the nuclear mass.

MATURE LIGNITE A descriptive term which refers to high-rank lignite, near sub-bituminous rank. Also see *Lignite*.

MAXIMUM PERMISSIBLE DOSE (MPD) The maximum radiation dose permitted by regulations for a particular set of circumstances. Maximum permissible dose depends on the exposed population, time period involved, type of radiation, and other factors. Also see *Lethal Dose*.

MBD, mbd, md/d Million United States barrels per day (oil); thousand barrels per day of NGL.

MBDO, mbdo Million barrels per day of oil.

mbdoe Million United States barrels per day of oil equivalent.

MBG Medium-Btu gas.

Mbm Thousand feet board measure.

MBPD Million barrels per day.

MCA Manufacturing Chemists Association.

Mcf, mcf Thousand cubic feet (gas); M represents 1000 in Roman nomenclature, not million.

mcfd Thousand cubic feet per day (gas).

MECHANICAL HYSTERESIS See Hysteresis.

MEDIUM-BTU GAS Also called intermediate-Btu gas as contrasted to low- and high-Btu gas. Medium-Btu gas, such as coal gas, coke gas, and synthesis gas, has a heat value of 13 to 26 MJ/m^3 (350 to 700 Btu/ft^3). Carbon monoxide, hydrogen and sulfur gases may be included as well as hydrocarbon gases. Synthesis gas may be upgraded to pipeline quality. Also see *Coal Gas; High-Btu Gas; Lean Gas; Low-Btu Gas.*

MEDIUM VOLATILE BITUMINOUS COAL Bituminous coal of the nonagglomerating type which has 69 to 78% fixed carbon (dmmf), and 22 to 31% volatile matter (dmmf).

MEGATON (MT) A term describing the energy release in a nuclear explosion. A one megaton nuclear explosion releases the same amount of energy as one million tonnes of TNT (at 1000 cal/g), or 4.18×10^{15} J. Also see *Kiloton.*

MEGAWATT Represented by MW, meaning 1,000,000 watts or 1,000 kilowatts of energy.

MERCAPTANS Organic compounds in the gasoline fraction of refined petroleum which include sulfur and give off an offensive "sulfur" odor. Mercaptans may be added in small amounts to natural gas to alert people of leaks in the system.

MERES Matrix of environmental residuals for energy systems (BNL).

MESH A measure of the number of openings in a screen or sieve designated as openings or meshes per inch or meshes per square inch (Table M-1). A 10 mesh per inch screen is 100 meshes per square inch. The actual size of the openings depends on the dimension of the wire or thread from which the screen or sieve is made.

TABLE M-1
Openings in U.S. Standard Sieves

Mesh no. meshes/inch	Nominal sieve opening mm	in
4	4.76	0.187
6	3.36	0.132
8	2.38	0.0937
12	1.68	0.0661
16	1.19	0.0469
20	0.841	0.0331
30	0.595	0.0234
50	0.297	0.0117
100	0.149	0.0059
200	0.074	0.0027
270	0.053	0.0021

MESON An elementary particle, positive or negative or neutral, which is unstable with a short half-life, associated with the nucleus. The lightest meson (pi) has about 250 times the mass of an electron.

MESOTHERMAL Hydrothermal deposits formed geologically which are sources of energy formed at medium temperature. Also see *Geothermal*.

MESOZOIC See Era.

METABOLIZABLE ENERGY Refers to biological energy relationships. Metabolizable energy is the gross energy of food consumed minus fecal energy minus energy in the gaseous products of digestion minus energy in the urine.

METAL HYDRIDE See Hydride; Natural Gas Hydrate.

METALLURGICAL COAL Coal suitable for use to produce a metallurgical coking coal which is characterized by low ash, low sulfur, low phosphorous, and high fixed carbon; appropriate carbonization properties; and certain physical characteristics. Several coals may be blended to provide a desirable coke for metallurgical processes, particularly for blast furnaces.

METALS Materials characterized by high electrical conductivity. Metals have wide application in energy production and use as structural materials, electrical and heat conductors, etc. The mechanical properties of most metals are affected by neutron irradiation. For this reason, and also because many metals absorb neutrons, there are only a few metals, like zirconium, that are well suited to use in the core of a nuclear reactor.

METER (or Metre) A unit of length measurement in the metric system. The basic unit of length for the SI units is defined as 1,553,164.13 wavelengths of the red line of cadmium in air at 760 mm pressure and 0°C. A yard is 3600/3937 meters. A meter is 39.37 in.

METHALINE An engine fuel consisting of 10% methanol and 90% gasoline.

METHANATION A process for producing methane by passing carbon oxides produced from gasified coal or oil and hydrogen mixtures over nickel catalysts. A medium-Btu gas can be upgraded to high-Btu gas by methanation. The reaction is

$$CO + 3\ H_2 \rightarrow CH_4 + H_2O;\quad CO_2 + 4\ H_2 \rightarrow CH_4 + 2\ H_2O$$

Also see *High-Btu Gas; Medium-Btu Gas*.

METHANE A colorless high-Btu gas (1000 Btu/ft) consisting of carbon and hydrogen, CH_4, which burns with a pale flame and has a low heating value of 50,000 kJ/kg (21,500 Btu/lb). The weight relative to air is 0.555. Natural gas contains at least 80% methane. Marsh gas is predominantly methane. Methane can be obtained from coal, oil, and biomass. Methane is combustible with air and is explosive in a concentration between 5.3 and 14% in air. Also see *Beninum; Biogas; Fiery Gas; Firedamp; Marsh Gas; Natural Gas*.

METHANOL An alcohol, also known as methyl alcohol or wood alcohol, the simplest alcohol, which has a formula of CH_3OH. Originally made by distillation from wood, methanol is now usually made from natural gas. Methanol is poisonous and cannot be used as a beverage. Methanol can be mixed with gasoline or used separately as a fuel. Also see *Absolute Alcohol; Ethanol; Gasohol.*

METHLINE A mixture of methanol and gasoline, for example, Methline 10 is 10% methanol. Also see *Methaline.*

METHOGENIC A substance which will produce methane.

METHOL A primary alcohol especially methyl alcohol. Also see *Methanol.*

METHYL TERTIARY BUTYL ETHER (MTBE) An additive for unleaded gasoline used to increase the octane number made from methanol, natural gas, and isobutylene.

METRIC SYSTEM A system of measurement based on the meter for length and gram or kilogram for weight. The meter and gram are subdivided or multiplied in decimal form. Time is measured in seconds. The metric system is represented by the CGS (or cgs), centimeter-gram-second, or MKS (or mks), meter-kilogram-second, designations. Recently the SI (International System) has been adopted by many groups. Also see *Conversion Factors; SI Units.*

MeV Million electron volts.

mf Moisture free.

MGD Million gallon per day.

MHD Magnetohydrodynamics.

MICROCURIE A millionth of a curie, 10^{-6} curie, a measure of radioactivity in terms of disintegrations per second. One curie is 3.7 x 10^{10} disintegrations per second. Also see *Curie; Millicurie.*

MICRON A unit of length equal to 0.000,001 m.

MICROPROCESSOR A microcomputer dedicated to a single task which is usually not reprogrammed; the unit has a large-scale integrator and provides a general-purpose data source for connection with external devices.

MILL As used in electric system cost and price analysis and tax rate, one-tenth of a cent (0.1¢) or ($0.001).

MIGRATION The movement of ions in an electrolyte during electrolysis; the movement of oil and gas through porous and permeable rock; the movement of water through soil or porous media.

MILLIARD A term for a large number, abbreviated m or ma, which represents 1,000 million, or 10^9.

MILLICURIE A thousandth of a curie, 10^{-3} curie, a measure of radioactivity in terms of disintegrations per second. One curie is 3.7×10^{10} disintegrations per second. Also see *Curie; Microcurie.*

MINE DISASTER See Major/Minor Mine Disaster.

MINERAL CHARCOAL See Mother of Coal.

MINERAL OIL See Oil.

MIUS Modular integrated utility system.

MIXING CHAMBER A compartment preceding a burner in which gas and air or other gases are mixed to provide correct ratio for burning. Also see *Manifold.*

M, m Thousand for gas, million for other applications.

MM, mm Million for gas.

MMBD Million barrels per day.

MM Btu Million Btu in gas.

m, mmf Moisture, mineral matter-free.

mmcf, MMCF 1,000,000 ft^3 of gas.

mmf Mineral matter-free.

mmscf, MMSCF 1,000,000 standard ft^3 of gas (Note: For gas M = 1000 SCF).

MOBILIZATION FOR SURVIVAL (MFS) A coalition of individuals promoting an antinuclear movement, located in Philadelphia, PA 19104.

MODERATOR A material such as graphite or heavy water, D_2O, placed in the core of a nuclear reactor to slow down the fast neutrons. The objective of slowing down the fast neutrons is to increase the probability of fission. Physical characteristics of good moderators for nuclear power plants include low atomic weight, high atomic density, good stability from heat and radiation, and good heat transfer properties. Also see *Graphite; Heavy Water.*

MOHOLE The name given to a hole, never yet successfully drilled through the crust into the mantle of the earth to obtain scientific information. The name mohole is derived from the name given to the discontinuity which exists between the crust and mantle, Mohorovicic. Also see *Earth; Mantle.*

MOISTURE CONTENT The amount of moisture in a material. (a) Several classifications or descriptions of moisture content are used, such as bound moisture, unbound moisture, equilibrium moisture, free moisture, capillary moisture, gravitational moisture, and osmotically held moisture (Figure M-2). The relationship of weight of moisture in a product to the weight of product is often expressed as a percent. Several terms are used to describe moisture content. Expressed as dry basis:

$$\text{M.C., \%d.b.} = \frac{\text{weight of water in product}}{\text{weight of dry matter}} \times 100\%$$

Expressed as a wet basis,

$$\text{M.C., \%w.b.} = \frac{\text{weight of water in product}}{\text{weight of dry matter + weight of water}} \times 100\%$$

The M.C., %w.b. and %d.b. are related as,

$$\text{M.C., \%d.b.} = \frac{\text{moisture content, \%w.b.}}{\text{100 - moisture content, \%w.b.}}$$

The wet basis of moisture content representation is used in commercial trade for fuel, food, organic and inorganic aggregates.

MOLECULAR WEIGHT The sum of the atomic weights of the atoms in a molecule of an element or compound. Also see *Gram Molecular Weight.*

MOLECULE The smallest unit of matter which can exist by itself and which retains the properties of the original material.

MOLLIER CHART The name applied to a method of representing graphically the thermodynamic properties of a pure substance with the enthalpy, h, on the ordinate and the entropy, s, on the abscissa. Other properties represented on the graph include pressure, superheat, moisture, and temperature. Also see *i-x Diagram.*

MON Motor Octane Number.

MOTHER COAL A soft coal rich in carbon which has a satin luster, which contains the cell structure of plants which can be seen under a microscope.

MOTHER OF COAL A lusterless coal, high in carbon, with plant structure visible, which occurs in thin layers in bituminous coal. Also called fusain and mineral charcoal. Also see *Fusain.*

MOTOR ENERGY SAVINGS The energy saved during motor operation as a result of various conservation measures. The energy used by 3-phase induction motors depends on size of motor, percent of full load, and starting periodicity. Devices are now becoming available to control the starting voltage at a low level and to reduce the voltage to a

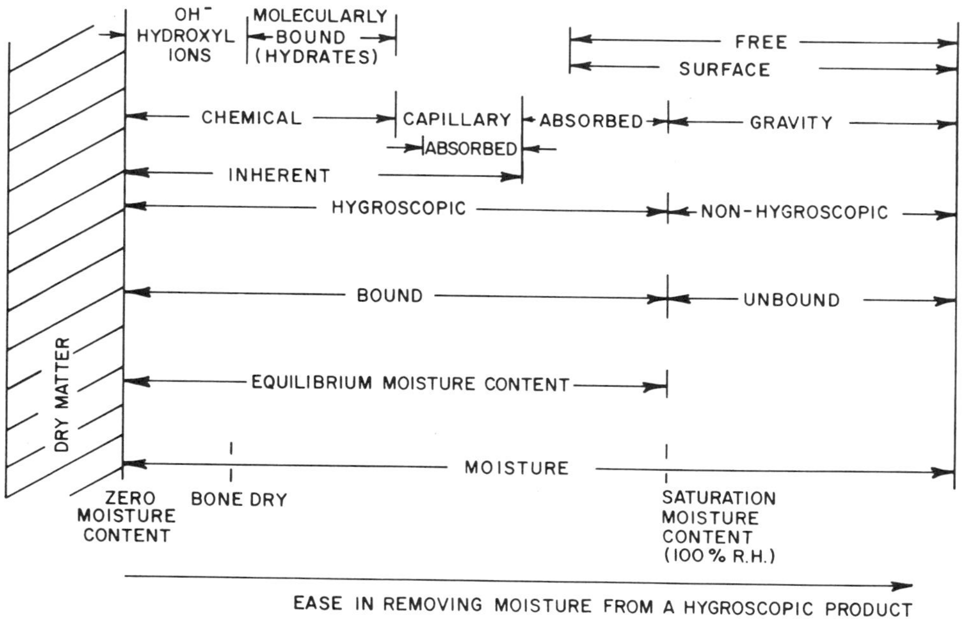

Fig. M-2 Moisture content representation. (Hall, 1979)

motor operating at less than full load. Energy savings may be as much as 50% for some applications.

MPD Magnetoplasmadynamic; maximum permissible dose (radiation).

MPG, mpg Miles per gallon.

MPH, mph Miles per hour.

MPM, mpm Meters per minute.

MPS, mps Meters per second.

MRG Methane-rich gas.

MRI Midwest Research Institute.

mscf, MSCF 1,000 standard cubic feet of gas.

MSR Molten salt reactor. (See Circulating-Fuel Reactor.)

MSW Municipal solid wastes.

MT Megatons.

mta Million tonnes per annum.

MTBE Methyl tertiary butyl ether.

mtce Million tonnes or tons of coal equivalent. One million tons coal equivalent is approximately equal to 0.6 million tonnes of oil.

mtoe Million tonnes or tons of oil equivalent; one million tons oil equivalent is equal to 1.67 million tons of coal.

MUD The general name of the drilling fluid used for gas or oil wells consisting of fine silt and clay particulates.

MUNICIPAL SOLID WASTES (MSW) The materials collected from urban areas in the form of organic matter, glass, plastics, waste paper, etc., not including human wastes. These materials can be used as a solid fuel for energy or converted to gaseous or liquid fuels (Figure M-3). Municipal solid wastes contain on the average in the U.S.A. (EPA) 53% paper, 8% organic scraps, 24% rags and household items, and 15% glass and metals. About 12,000 kJ/kg (5300 Btu/lb) can be recovered from MSW, or 10.5 million kJ/ton. Also see *Volatile Solid*.

MUON A nuclear particle, also called the mu meson, which has a rest mass equal to 207 times the electron rest mass. Both positive and negative muons exist. The muon is an unstable particle found as one component of cosmic rays.

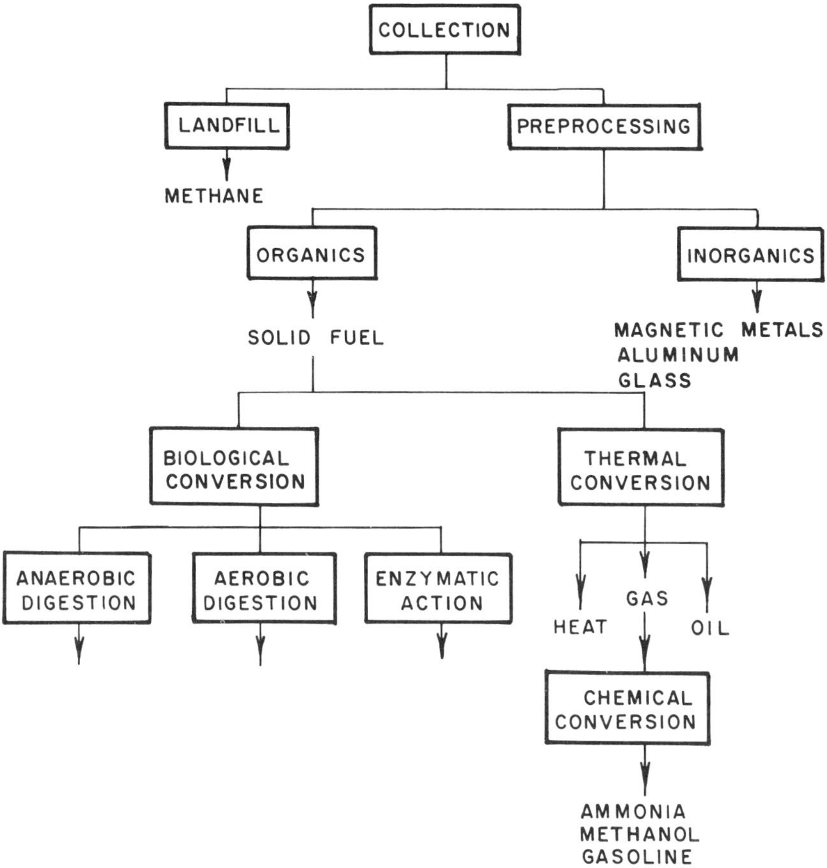

Fig. M-3 Municipal refuse options *(Bailey, 1978)*.

MW Megawatt.

MWe Megawatt-electrical.

MWP Maximum working pressure.

MWth Megawatt-thermal.

N

NAAQS National Ambient Air Quality Standards (Clean Air Act, PL95-95).

NAE National Academy of Engineering.

NAK, NaK A mixture of sodium and potassium (liquid). See Liquid Metal.

NAPHTHA A light distillate made from refining petroleum which is removed in the 70°-200°C (158°-392°F) range. Naphtha is used for blending with other components to make gasoline, as a feedstock to make chemicals, and to make synthetic natural gas (with steam at 700°C (1290°F) at 3 atm). Naphtha bottoms refer to the residue in the steam after removing the naphtha by distillation. Light naphtha is used as a solvent in the manufacture of paint and varnish. Also see *Distillate; Tar Oils.*

NAPHTHENE SERIES (Cycloparaffins) Saturated hydrocarbons which are similar to paraffins in properties except that the carbon atoms for each molecule are arranged in a closed ring. Naphthenes can be made into higher octane aromatics by catalytic dehydrogenation. The naphthenes occur in the residues of high-boiling point 400°C (750°F) petroleum. Also see *Naphtha; Hydrocarbon Series.*

NARUC National Association of Regulatory Utility Commissioners.

NAS National Academy of Sciences.

NASA National Aeronautics and Space Agency.

NATIONAL ASSOCIATION OF POWER ENGINEERS 176 W. Adams Street, Suite 1914, Chicago, IL 60603.

NATIONAL ENERGY ACT The name given to a body of energy legislation passed by the United States Congress in 1978. The legislation covered the following subjects: Encouragement of energy conservation, natural gas price deregulation, substitution of coal for petroleum and natural gas, utility rate reform, and tax incentives for energy-saving investments.

NATIONAL FIRE PROTECTION ASSOCIATION 470 Atlantic Avenue, Boston, MA 02110.

NATIONAL PETROLEUM COUNCIL (NPC) An advisory group to the secretary of the United States Department of Interior on petroleum and gas subjects, located at 1625 K Street, N.W., Washington, DC 20006. Also see *American Petroleum Institute; Economic Regulatory Administration.*

NATIONAL RURAL ELECTRIC COOPERATIVE ASSOCIATION 1800 Massachusetts Avenue, N.W., Washington, DC 20009.

NATIVE The adjective used to identify a substance as found in nature, or as associated with a particular location, such as native asphalt, native bitumen, native coke, native gas, etc., in contrast to substances manufactured or produced artificially. Also see *Natural.*

NATIVE GAS Gas originally present in coal, oil, or as a gas in a reservoir in the ground.

NATURAL An adjective applied to the form, shape, and make-up of substances found in nature, such as natural asphalt (before crushing), natural carbon (before modification), natural coke (made by natural processes), natural gas (as found in nature), etc. Also see *Native.*

NATURAL GAS A hydrocarbon gas as it comes from the well, consisting principally of methane (85 to 95%), which may continue to be called natural gas after being treated to remove some of the components. Gas removed from crude oil is associated gas. Gas with more than 5 g/m^3 of hydrogen sulfide is sour gas. If the condensate is removed it may be called stripped gas. The range of composition of natural gas, depending on location of source, varies greatly. Typical values of the percentage composition of hydrocarbons in natural gas are as follows:

	Wet	Dry
Methane	84.6	96
Ethane	6.4	2.0
Propane	5.3	0.6
Butane	2.6	0.30

Other components may include from 0 to 5% of carbon dioxide, helium, hydrogen sulfide, and nitrogen. The calorific value is usually in the range of 33 to 41 MJ/m^3 (900-1100 Btu/ft^3). Natural gas can be liquefied by cooling below -46°C (-49°F) during which the volume change is 635 to 1. Also see *Butane; Ethane; Liquefied Natural Gas; Methane; Raw Gas.*

NATURAL GAS HYDRATE A naturally occurring material arranged as a lattice with the host being water with molecules of methane attached. One cubic foot of hydrate yields approximately 170 scf of methane (natural gas).

NATURAL GAS LIQUIDS Hydrocarbons, such as propane and butane, which are produced as part of the product of natural gas wells but can be separated from the gas and maintained as liquids. They may be removed from natural gas with separators, scrubbers, and gas processing plants. Also see *Liquefied Natural Gas; Natural Gas.*

NATURAL RADIOACTIVITY Radioactivity of materials found in nature. Much of this natural radioactivity is produced by the decay of thorium and uranium and their daughter products. Uranium and thorium were produced in the early days of the universe. On the other hand, some natural radioactive materials, like ^{14}C, are produced continually by cosmic rays. Also see *Daughter Products.*

NBS National Bureau of Standards.

NCA National Coal Administration.

NEAT ALCOHOL Alcohol processed until the product obtained is 100% alcohol or an absolute alcohol. Neat alcohol is 200 proof.

NEB National Energy Board (Canada).

NEGENTROPY A quantity derived from entropy by changing the algebraic sign. Thus if the entropy of a system is S, the negentropy is -S. The change in sign means that the negentropy of a system is larger the greater its degree of organization.

NEIC National Energy Information Center.

NEMA National Electrical Manufacturers Association.

Neo- A prefix meaning new or recent, opposite of paleo-.

NEP National Energy Plan.

NEPA National Environmental Policy Act.

NEPCO New England Petroleum Corporation.

NER Net energy ratio.

NERA National Energy Resources Association.

NERC National Electric Reliability Council.

NET ENERGY The amount of energy delivered to or used at the end point not considering inefficiencies, losses, and waste previously incurred. Also see *Gross Energy; Gross Heating Value; Net Heating Value.*

NET ENERGY RATIO (NER) The energy output of an energy conversion system divided by the energy input to the system. For sugar cane to ethanol, 4.53 units of energy are produced per unit of commercial energy input; thus, a NER of 4.53, which reflects the high input of solar energy to the production.

NET HEATING VALUE See Low Heating Value.

NEUTRINO An uncharged particle that is created during radioactive decay. There are at least two and possibly three types of neutrinos, each of which has a related anti-neutrino. Neutrinos and anti-neutrinos interact very weakly with matter.

NEUTRON An uncharged nuclear particle. The fissions that produce the energy in a nuclear reactor are initiated by neutron absorption in the nuclei of fuel atoms. The free neutron is radioactive.

NEUTRON ACTIVATION The process by which a radioactive nucleus is produced from a stable nucleus as a result of the absorption of a neutron. The characteristic radiation from the radioactive decay of the product nucleus can be used to detect the presence of the original nucleus. Thus, neutron activation is a method of chemical analysis. Also see *Induced Radioactivity.*

NEUTRON CAPTURE A nuclear reaction during which a neutron enters and is captured by a nucleus. Following the capture, gamma rays are usually emitted by the new nucleus formed. Neutron capture is an alternative to nuclear fission following the entrance of a neutron into a fissile nucleus. The neutron capture process usually produces a radioactive nucleus.

NEUTRON DENSITY The number of neutrons per unit volume. In a nuclear reactor the phrase is used to describe the density of free neutrons.

NEUTRON FLUX The product of neutron density and neutron speed. The neutron flux is directly used in calculating nuclear reaction rates.

NEUTRON NUMBER An integer (N), equal to the number of neutrons in a nucleus. It is related to the mass number (A) and charge number (Z) by the equation $N = A - Z$.

NEWTON A unit of force. One newton is that force which will give 1 kg_m (mass) an acceleration of 1 m/sec^2. One newton is equal to 10^5 dynes. Also see *Force; Joule; Work.*

nfe Nitrogen-free extract.

NFPA National Fire Protection Association (see Fire Classification).

NGL Natural gas liquids.

NGPA National Gas Processors Association.

NIOSH National Institute for Occupational Safety and Health.

NITROGEN A gaseous element, N, relatively inert, which makes up 78% of the atmosphere. Although nitrogen is a part of the air taken in during the combustion process, it doesn't contribute energy; depending on the circumstances of combustion, particularly at high temperature, nitrogen will form nitrogen oxides which are pollutants in the environment. The properties of nitrogen are: Atomic weight, 14.007; atomic number, 7; boiling point, -195.8°C (-321°F); melting point, -209.90°C (-346°F); density, 1.25 g/liter; and specific gravity of liquid, 0.808. Also see *Liquid Nitrogen; Nitrogen Oxides*.

NITROGEN/CARBON RATIO The ratio of nitrogen to carbon in a substance, usually on an atomic basis. Examples of N/C atomic ratio are: Oil shale (Colorado), 0.028; Athabasca tar sand (Canada), 0.004; bituminous coal, 0.018; lignite, 0.015; and anthracite coal, 0.001. Also see *Carbon/Nitrogen Ratio*.

NITROGEN OXIDES The oxides of nitrogen (NO_x), such as those produced from nitrogen in the air during combustion. The federal standard for automibile emission in 1981 was 1 gpm (gram per mile). The approximate production of oxides of nitrogen from burning are as follow (*Junge, 1980*):

Fuel	Lb/MMBtu	Ng/J
Gaseous fuel	0.20	86
Liquid fuel	0.30	130
Wood	0.32	138
Solid fossil fuel	0.70	300

NITS National Information and Technical Service.

NLGI National Lubricating Grease Institute.

NMSS Nuclear Materials Safety and Safeguards.

NOAA National Oceanic and Atmospheric Administration.

NOBLE GAS Those gases which are nearly inert and do not react with other chemicals. Also known as inert gases and rare gases. The noble gases are argon (Ar), helium (He), Krypton (Kr), Neon (Ne), and Radon (Rn). Also see *Krypton; Radon*.

NON-ASSOCIATED GAS A dry hydrocarbon gas which is not associated with oil in a reservoir which is producing oil. Also see *Associated Gas*.

NONIONIZING RADIATION Radiation which is normally not capable of dissociating atoms or molecules into charged particles.

NORMAL ATMOSPHERE The ambient atmospheric conditions at 1 atmosphere pressure (760 mm), 0°C (32°F), at a latitude of 45°. Also referred to as NTP, normal temperature and pressure. Also see *NTP*.

NO_x Nitrogen oxides (see Nitrogen).

Np Neptunium.

NPC National Petroleum Council.

NPCC Northeast Power Coordinating Committee.

NPRA National Petroleum Refiners Association.

NRC Nuclear Regulatory Commission (formerly part of the AEC); also, National Research Council (National Academy of Sciences).

NRECA National Rural Electric Cooperative Association.

NRSE New and renewable sources of energy (UN).

NSF National Science Foundation.

NSPAC National Standard Policy Advisory Committee.

NSPS New source performance standards.

NSSS Nuclear steam supply system.

NTP or STP Normal temperature and pressure or standard temperature and pressure of a gas for standard, usually taken as 0°C (32°F) or 20°C (68°F) or sometimes 25°C (77°F), and a pressure of 1 atmosphere (760 mm). Normal or standard conditions are often used in making calculations and comparisons. Also see *Normal Atmosphere; STP*.

NUCLEAR BATTERY A battery that utilizes the thermoelectric effect to generate a voltage in which the heat source required to generate a thermoelectric potential is supplied by a radioisotope incorporated in the battery.

NUCLEAR BINDING ENERGY The difference between the total rest energy of the protons and neutrons that make up a nucleus and the total rest energy of the nucleus itself. The rest energies of the nuclei and particles are related to their masses by the Einstein formula $E = mc^2$.

NUCLEAR CROSS SECTION An apparent cross sectional area presented by a nucleus to an impinging beam of particles. The cross section depends on the type of nuclear reaction involved, kind and energy of impinging particles, and other variables. Cross sections are measured in area units.

NUCLEAR ENERGY; ATOMIC ENERGY A term referring to energy released during the conversion of nuclear rest mass into energy.

NUCLEAR FISSION A nuclear reaction during which a nucleus divides into two or, on rare occasions, three major fragments heavier than helium nuclei. The reaction may be spontaneous or induced by an impinging particle such as a neutron.

NUCLEAR FUEL CYCLE The sequence of processes involved in fuel preparation, energy extraction and waste disposal for a nuclear fuel, as represented by Figure N-1.

NUCLEAR FUELS Materials that are used as practical sources of nuclear energy. Usually a nuclear fuel consists of an isotope that releases nuclear energy in a fission or fusion reaction plus other elements that provide a matrix suitable for use in a fission or fusion reactor. The most common nuclear fuel for fission is uranium dioxide made with uranium enriched in the isotope ^{235}U, the fuel used in light water reactors. The principal nuclear fusion fuels are ^{2}H (deuterium) and ^{3}H (tritium).

Nuclide	Nuclear Reaction (releasing energy)
Fission	
^{233}U	$n + {}^{233}U \rightarrow$ fission
^{235}U	$n + {}^{235}U \rightarrow$ fission
^{239}Pu	$n + {}^{239}Pu \rightarrow$ fission
Fusion	
$^{2}H_1$	${}^{2}H_1 + {}^{2}H_1 \rightarrow {}^{3}He_2 + n$; ${}^{2}H_1 + {}^{2}H_1 \rightarrow {}^{3}H_1 + p$

Sometimes the term nuclear fuel is used to denote a mixture of fissile isotopes and other materials taken together. Thus, for example, in a light water reactor, the pellets of uranium oxide including both $^{235}UO_2$ and $^{238}UO_2$ are usually called the nuclear fuel.

NUCLEAR INSURANCE Liability insurance protecting the owner of a nuclear reactor or other nuclear facility from claims for damages. At present an industry association provides limited insurance to the nuclear industry. Under the Price-Anderson Act the federal government supplies coverage from the insurance association limit up to a higher

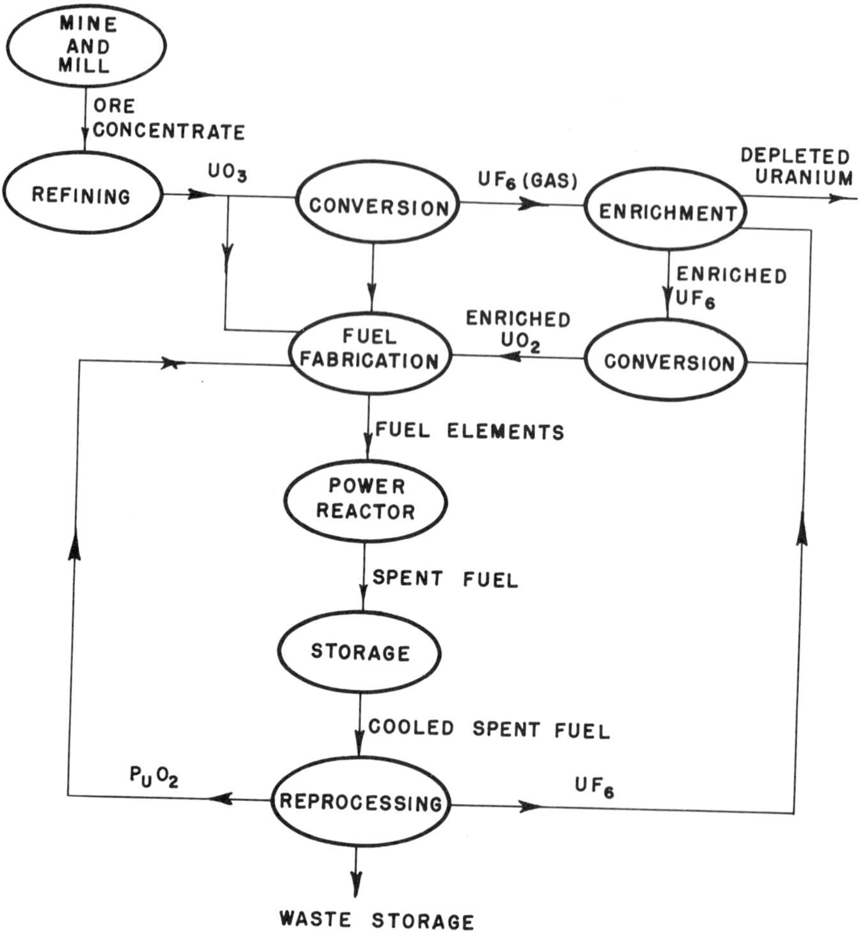

Fig. N-1 Nuclear fuel cycle *(Schultz and Whicker, 1980; IAEA, 1976)*.

limit. There is no coverage beyond the government limit. This insurance arrangement is temporary and may well be altered in the future. Also see *Price-Anderson*.

NUCLEAR POWER PLANT An electrical generating plant which includes a nuclear reactor as the heat source. Almost all United States nuclear power plants utilize either a pressurized water reactor (PWR) or a boiling water reactor (BWR). Gas-cooled and heavy water reactors are common in other countries.

NUCLEAR REACTION A reaction that involves interaction between an incident particle and a nucleus through nuclear forces.

NUCLEAR REACTOR Equipment used to sustain a large scale neutron chain reaction and thus to produce large neutron fluxes and heat from nuclear fission. Research and test reactors primarily use the neutrons. Power reactors use the heat. Types of power reactors in common use in the United States include pressurized water reactors and boiling water reactors (Table N-1).

TABLE N-1
Nuclear Reactors in Operation, June 30, 1980

	World		United States	
Reactor type	Number	Net, MWE	Number	Net, MWE
Pressurized water	91	61,641	41	31,839
Boiling water	58	36,185	26	17,813
Pressurized heavy water	13	5,940	0	0
Gas cooled	44	11,242	1	330
Other thermal	19	10,052	1	860
Fast	5	1,493	1	60

Source: *Nuclear News, Aug. 1980.*

NUCLEAR REACTOR CORE The portion of a nuclear reactor where the nuclear fuel is located and where nuclear fission takes place.

NUCLEAR REACTOR WASTE The radioactive and non-radioactive materials that are produced as waste products of the operation of a nuclear reactor. Radioactive wastes include fission products, unused fuel, neutron activation products, and transuranic elements.

NUCLEAR REACTOR SAFETY The topic that covers various aspects of protecting workers and the public from injury that might result from the operation of nuclear reactors.

NUCLEAR REGULATORY COMMISSION (NRC) A group of five commissioners established by law in 1974, as a successor of the Atomic Energy Commission (AEC), which has federal authority and responsibility for licensing and regulating the safety and environmental impacts of nuclear power plants and related activities. Activities of the Commission include siting, licensing, developing standards for operation, supervision, fuel handling, and waste disposal. Also see *Department of Energy; Economic Regulatory Administration; Federal Energy Regulating Commission.*

NUCLEAR SITING Selecting the site for a nuclear power plant. The process includes consideration of such factors at a proposed site as proximity to electric power markets and transmission corridors,

TABLE N-2
Number Prefixes for SI and Metric Units

Multiples and Submultiples	Prefixes	Symbols
10^{18}	exa*	E
10^{15}	peta	P
10^{12}	tera	T
10^{9}	giga**	G
10^{6}	mega	M
10^{3}	kilo	k
10^{2}	hecto	h
10	deka	da
10^{-1}	deci	d
10^{-2}	centi	c
10^{-3}	milli	m
10^{-6}	micro	μ
10^{-9}	nano	n
10^{-12}	pico***	p
10^{-15}	femto	f
10^{-18}	atto	a

*Also giga giga, GG
**Also bera, B
***Also micromicron, μμ

availability of cooling water, geological stability, population density, and availability of a construction site with appropriate land contours and soils.

NUCLEAR WASTES The wastes from commercial nuclear power operations and use of radionuclides may be classified as low-level, high-level, and transuranic, which may be gaseous, liquid, or solid. Also see *High-Level Wastes (HLW); Low-Level Radioactive Wastes (LLRW); Nuclear Reactor Waste; Transuranic Wastes.*

NUCLEON A neutron or proton. This name is meant to include both types of particles that make up nuclei.

NUCLEONICS The subject that covers the science and technology of nuclear reactions and interactions.

NUCLIDE A species of atom characterized by the number of protons and neutrons in the nucleus. Radioactive nuclides make up a sub group of all nuclides. Also see *Radioactive Nuclide; Radionuclide.*

NUMBER OF NUCLEAR POWER PLANTS See Nuclear Reactor.

NUMBER PREFIXES Names placed before words to represent large to small quantities (Table N-2). The names representing large numbers, 10^6 and above, have as a symbol capital letters, such as G for giga. The names representing numbers of 10^3 and less are represented by small letters, such as k for kilo, as shown in Table N-2.

NURE Nuclear Uranium Researches Estimate (program).

NUTATION The heliotropic movement of a sunflower in which the head turns toward the sun during the day. The movement results from bending of the stem during the day, straightening during the night, so the flower faces east in the morning. *(Sci. Am. 244(5):150)*. Nutation also refers to the changing orientation of the axis of rotation of a top.

O

O_X Total oxidants.

OAPEC Organization of Arab Petroleum Exporting Countries.

OBR Oxygen blown reactor.

OBV Octane blending values.

O/C Oxygen/carbon.

OCEAN CURRENTS The movement of water in the ocean from which energy could be recovered, as a result of temperature gradients, density gradients, wind, and tides. Also see *Ocean Thermal Energy Conversion (OTEC); Wave Power.*

OCEAN THERMAL ENERGY CONVERSION (OTEC) A procedure for using the temperature differences, called ocean thermal gradient, in the ocean to generate electric power. The system consists of an engine that uses warm ocean surface waters as a heat source and rejects waste heat to the cold water at greater depth in the ocean. The maximum temperature differences are found along the equator, where a temperature of 30°C (86°F) on the surface and 5°C (41°F) at a depth of 1000 m exist. To be economical, a temperature difference of 16.7°C (30°F) is needed. The systems are about 2% efficient. The energy in the ocean water could be used for producing steam at pressures in an open cycle or for a gas/liquid as the heating medium in a closed cycle. Also see *Ocean Currents; Tidal Energy.*

OCEAN THERMAL GRADIENT The variation of the temperature of ocean water with depth below the surface. Below a surface mixed layer there is a decline in ocean water temperature with increasing depth. The temperature difference between surface water and water at depth can, in principle, be utilized to operate an engine. Also see *Ocean Thermal Energy Conversion.*

OCR Office of Coal Research.

OCS Outer continental shelf.

OCTANE One of the group of linear chain hydrocarbons called alkanes. Octane has a formula of C_8H_{18}, naturally found in petroleum. It is one of a group of isomeric hydrocarbons, the best known of which is isooctane, a member of the gasoline family. In isooctane the atoms are arranged in the shape of a spider web. Octane is a colorless liquid which boils at 124.6°C (256°F), isooctane boils at 99°C (210°F). Both have a density of 0.7 g/ml, with octane very slightly less dense than isooctane. Also see *Isomer; Isooctane; Octane Number.*

OCTANE NUMBER A number representing the relative anti-knock value of a fuel in an internal combustion engine compared to the standard fuels isooctane (which has an octane number of 100) and heptane (which has an octane number of 0). The actual value of the octane number for a fuel is obtained in a CFR (Cooperative Fuels Research) engine which has an adjustable compression ratio. The octane number is also designated as RON, research octane number. Tetraethyl lead (TEL) or tetramethyl lead (TML) or methyl tertiary butyl ether (MTBE) may be added to gasoline in small quantities of 1.5 g or less per gallon to increase the octane number. Also see *Cetane Number; Diesel Index; Isooctane; Lead; Octane.*

OECD Organization for Economic Cooperation and Development.

OEP Office of Emergency Planning.

OFF-GAS The gas produced as a byproduct of pyrolysis of solids in which the major objective is usually to produce fuel oil, oil, and char. Also see *Refinery Gas.*

OIE Office of Inspection and Enforcement (NRC).

OIL The general term or name given to a viscous liquid at ordinary temperatures, insoluble in water and lighter than water, which possesses a lubricating quality. Oil may be in one of three categories, the first of which is most important as a fuel resource: (1) Mineral oils derived from fossil fuels, which consist principally of hydrocarbons; (2) fixed oils derived from animal, marine, or vegetable matter, which consist principally of glycerides and ester; and (3) essential oils derived from vegetable matter have characteristic odors. Fixed oils decompose and change chemical composition when heated. Mineral oils are also called crude oil and/or petroleum. Also see *Crude Oil; Petroleum.*

OIL ASPHALT Asphalt obtained from a petroleum source.

OIL COAL See Stellarite.

OIL FUEL Refined petroleum, crude petroleum, shale oil, tar sand,

and other substances from fossil liquids used as fuel.

OIL GAS A gas obtained by the destructive distillation of petroleum consisting principally of methane, ethylene, and acetylene.

OIL GAS TAR A heavy hydrocarbon produced by cracking oil vapors at high temperatures when making oil gas.

OIL GASIFICATION The process of producing gaseous fuels from a petroleum feedstock. a) From 50 to 90% of the carbon in the feedstock is usually converted to gas depending upon the operating conditions and the feedstock. The fluidized bed hydrogenation process is a noncatalytic reaction used for oil gasification. b) The hydrocracking-hydrogasification process developed by IGT involves first hydrocracking to produce an oil similar to diesel oil followed by gasification under pressure (500 to 1500 psi) at high temperature 590° to 760°C (1100° to 1400°F) in the presence of hydrogen. c) The Shell gasification process uses partial oxidation of oil in the presence of either air or oxygen producing a gas consisting primarily of carbon monoxide and hydrogen. Also see *Gasification.*

OIL IMPORTS The import of crude oil by the U.S.A. Oil imports increased rapidly in the 1970's, reaching a peak in 1977 (Table O-1).

TABLE O-1
U. S. Import of Crude Oil, million bbl/day (average)

Year	Amount
1950	1.2
1960	2.9
1970	5.0
1975	6.1
1977	8.8
1979	8.4
1980	5.2

OIL RECOVERY A term used to describe processes for extracting crude oil located in porous media in the crust of the earth. From 10% to 80% of the crude oil can be removed from a reservoir. Various techniques are used to increase the recovery on an economical basis through secondary and tertiary recovery techniques. Also see *Oil Recycling; Oil Re-use; Oil Sand; Secondary Recovery; Tertiary Recovery; Waste Oil Recovery.*

OIL REFINING See Petroleum; Refining.

OIL SAND The name given to the porous sandstone rock in which crude oil is located. Also see *Oil Recovery; Secondary Recovery.*

OIL SHALE A carbonaceous rock from which oil can be obtained upon heating to 430° to 540°C (800° to 1000°F). Kerogen in the oil shale is the precursor for the formation of oil. Oil shale can be converted into crude oil, gas, and carbonaceous residues by destructive distillation (Figure O-1). Considerable variation exists in the amount of oil which can be obtained from oil shale, being from 15 to 40 gal/ton. Also see *Carbonaceous; In Situ Combustion; Kerogen.*

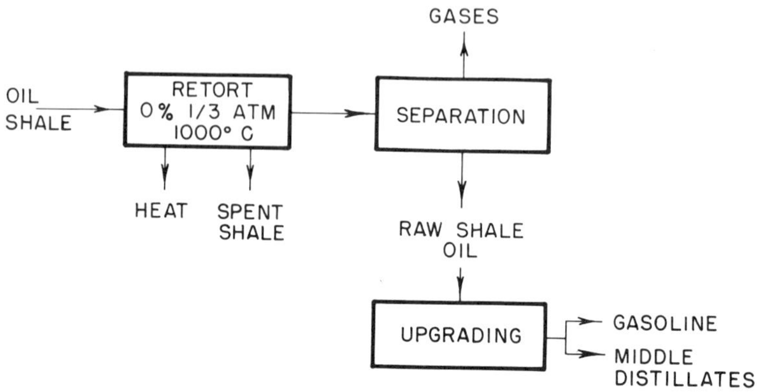

Fig. O-1 Oil shale retorting.

OIL SHALE RETORTING See Fushun Process.

OIL WELL GAS See Casinghead Gas.

OLEFINIC HYDROCARBONS Unsaturated hydrocarbons. If two hydrogen atoms can be added to an olefin, it is a mono-olefin; if four, a diolefin. Olefins are usually formed by cracking. The mono-olefins have names corresponding to those of the paraffins, which end in -ene or -ylene (ethylene). Diolefins have names corresponding to paraffins which end in -diene as in butadiene.

ONR Office of Naval Research.

OOG Office of Oil and Gas (USDI).

OPEC See Organization of Petroleum Exporting Countries.

OPEN CYCLE Operation of a heat engine or other power device in which the power fluid is used once, then replaced with fresh fluid, in contrast to a closed cycle in which the fluid is recirculated. An internal combustion engine is an open cycle. A refrigeration system is a closed cycle.

OPS Offshore power systems.

ORAU Oak Ridge Associated Universities.

ORDER OF MAGNITUDE A designation relating to the large difference between two numbers, with one number approximately 10 times another. Thus, the difference between 15 and 140 to 160 is an order of magnitude. Also see *Scale-Up of Equipment*.

ORGANIC Having the characteristics of material that is or once was living.

ORGANIC SHALE See Bituminous Shale.

ORGANIC WASTE See Municipal Solid Waste; Solid Waste.

ORGANIZATION OF PETROLEUM EXPORTING COUNTRIES (OPEC) An oil cartel founded in 1960. The members of OPEC (1981) are Algeria, Ecuador, Gabon, Indonesia, Iran, Iraq, Kuwait, Libya, Nigeria, Qatar, Saudi Arabia, United Arab Emirates (Abu Dhabi, Dubai, Sharjah), and Venezuela. The OPEC headquarters office is in Vienna, Austria. The principal aims of the organization are the coordination and unification of the petroleum policies of member countries and the determination of the best means for safeguarding their interests, individually and collectively.

ORMAK Oak Ridge Tokomak.

ORNL Oak Ridge National Laboratory.

OSHA Occupational Safety and Health Administration (and Act).

OTA Office of Technology Assessment.

OTEC Ocean thermal energy conversion.

OVERALL HEAT TRANSFER COEFFICIENT A quantity often represented by U, which characterizes the heat transfer through a particular thermal barrier (e.g., a wall) or one of its components (Table O-2). The units for U are: Heat units per unit area - temperature difference - time, W/m^2K or $Btu/hr\ ft^2\ °F$. $U = 1/R$, in which R is the overall resistance to heat transfer.

$$q = UA\ (t_1 - t_2)$$

where

U = overall heat transfer coefficient
A = area of heat transfer
$t_1 - t_2$ = temperature difference between the two sides of the thermal barrier or component.

TABLE O-2
Overall Heat Transfer Coefficients, U, Approximate Range of Values

General	Btu/hr ft² °F	W/m^2K
Freon condenser, water as a coolant	50-150	250- 800
Water-to-water heat exchanger	150-300	800-1650
Water-to-oil heat exchanger	20- 60	100- 350
Steam-to-water	200-700	1000-3900
Steam-to-gas	5- 50	25- 275
Types of dryers		
Vacuum shelf	1	5- 6
Indirect rotary	2- 10	10- 55
Jacketed trough	2- 15	10- 85
Rotary vacuum	5- 50	25- 275
Agitated pan	5- 60	25- 325
Drum	200-300	1150-1700

Conversion Factor: 1 Btu/hr ft^2 *°F = 5.678* W/m^2K

Also see *R-Value*.

OTTO CYCLE The sequence of events for an internal combustion engine for transforming energy in fuel to power on the shaft. A piston in a cylinder receives the impact of combustion of the fuel and transmits it to a shaft. Ignition is provided by an electric spark. Either two-stroke or four-stroke engines may be used (Figure O-2).

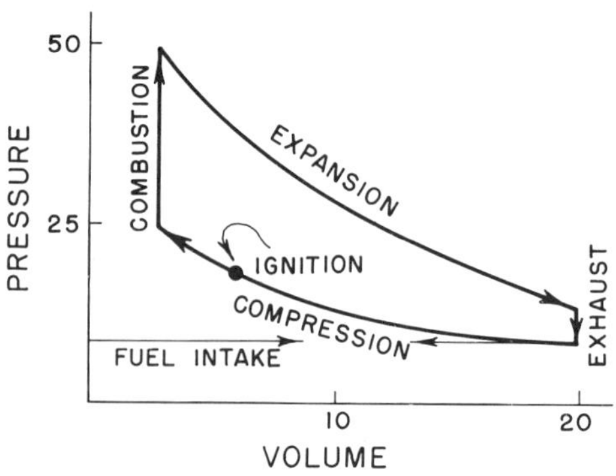

Fig. O-2 Otto cycle.

OVERALL THERMAL EFFICIENCY The ratio of the final useful energy output of a system to the heat energy (heat of combustion in the case of a fuel) input (Table O-3). Also see *Efficiency*.

TABLE O-3
Overall Thermal Efficiency of Units, Percent

Steam engine, thermal to mechanical	15
Internal combustion engine, gasoline	15-25
Diesel engine	27-34
Counterflow conveyor dryer	60-65
Rotary high temperature dryer	65-70
Boiler, steam	70-85
Heat exchanger, air to air	70-85
steam to fluid	70-80
water to water	80-90
Electric heating elements to liquids	over 90

OVERBURDEN The soil and rock between the surface of the earth and a coal or other ore seam.

OVERTHRUST BELTS Two geologically significant areas in the U.S.A., the Eastern Overthrust Belt which extends from Alabama to Canada, and the Rocky Mountain (Western) Overthrust Belt which extends from Mexico to Alaska (Figure O-3). These areas contain considerable reserves of oil and gas, which are now being intensively explored. In 1980 it was estimated that the Eastern Overbelt area contained 1.5 billion barrels and the Rocky Mountain Overthrust area contained 13 billion barrels of crude oil. Also see *Elephant Fields*.

OWPS Off-shore windpower system.

Oxy- Prefix pertaining to oxygen.

OXYGEN A colorless, odorless gas at NTP, which makes up 20.99% by volume of the air in the atmosphere, represented by O as an element and O_2 as a gas molecule. It is an element with atomic number 8, the common isotope has an atomic mass number 16, but stable isotopes also exist with mass numbers of 17 and 18. Oxygen is essential for the life of most organisms on earth; is needed for decomposition of refuse; and is an essential element in combustion. Oxygen is the most abundant element in the earth's crust, being approximately equal in weight to all other elements. Oxygen can be obtained from many sources, the most important of which are: Decomposition of water (electrolysis) separation from air (pressure/temperature) and oxygen-containing compounds (heating).

Physical properties of oxygen are: Boiling point, -182.96°C (-296.8°F); melting point, -218.4°C (-361.1°F); density, 1.429 g/L

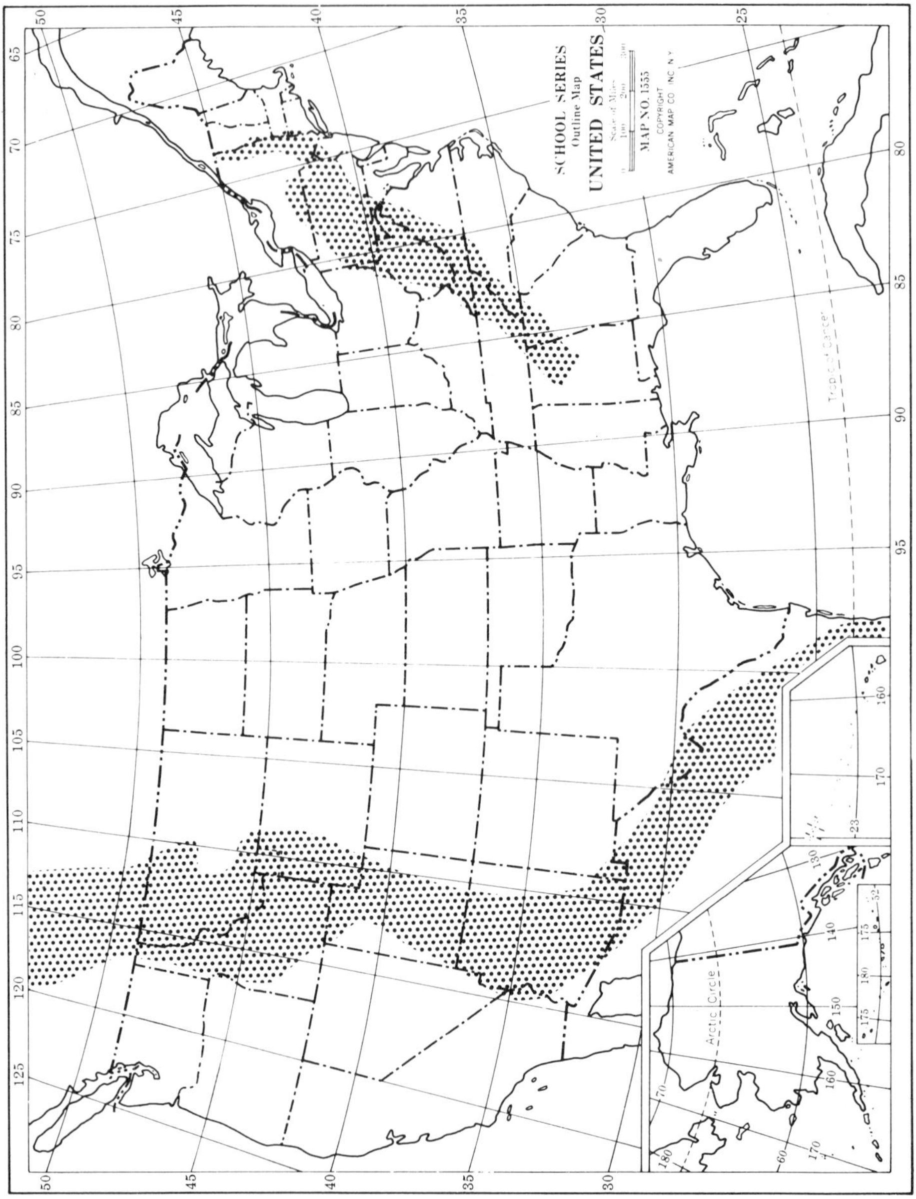

Fig. 0-3 Overthrust belts.

(at 0°C), and specific gravity of liquid, 1.14. Also see *Liquid Oxygen; Ozone*.

OXYGEN/CARBON (O/C) RATIO A parameter used in the process of coal conversion to a liquid, with the oxygen divided by carbon in moles. The maximum conversion of coal occurs at an O/C mole ratio of approximately 0.06. Also see *Carbon/Oxygen*.

OZONE An unstable gas, O_3, bluish in color with a distinctive odor, formed by an electric discharge in oxygen. Formed naturally in the atmosphere by lightning. There is a layer of ozone in the atmosphere at altitudes of 20,000 to 30,000 m (65,000 to 98,000 ft). Ozone is used for purification of air and water and as a bleaching agent. The properties of ozone: Boiling point, -112°C (-169.5°F); molecular weight, 24; density, 2.144 g/liter (1.5 times as heavy as oxygen), and when one gram of O_3 is converted to O_2, 717 cal of heat is liberated. Also see *Oxygen*.

P

PAD Petroleum Administration for Defense.

PAH Polycyclic aromatic hydrocarbons.

PALEOZOIC See Era.

PAN Peroxyacetyl nitrate (found in air pollution, smog).

PAR Photosynthetically active radiation. (See Photosynthesis.)

PARAFFIN GAS Gases which have a name ending in -ane, i.e., methane, butane, propane.

PARAFFIN HYDROCARBONS Saturated hydrocarbons in which the bonds of the carbon atoms are filled with hydrogen atoms. The saturated hydrocarbons are generally chemically inactive until broken apart.

PARAFFIN SERIES (Alkane) Hydrocarbons which contain the maximum number of hydrogen atoms, thus are saturated. These are chemically inactive under most conditions, but are chemically active when broken apart. The carbon atoms in the paraffin series form chains. The straight-chain structure (without branches) are normal paraffins; those with branches are isoparaffins. Paraffins with four or less carbon atoms are gaseous, those with five to 15 are liquids and those with over 15, waxes. The lighter gaseous paraffins, methane and ethane, are components of natural gas. The heavier gaseous paraffins, propane, butane and isobutane are components of liquefied petroleum gas (LPG). Also see *Aliphatic Hydrocarbons; Alkanes; Decane*.

PARTICLE ACCELERATORS Devices for accelerating sub-atomic particles—electrons, protons, photons, etc.,—for the purpose of studying the basic properties of matter and for applications such as medicine, radiation processing, and non-destructive testing (Table P-1). Also see *Billion Electron Volts (BeV)*.

TABLE P-1
Particle Accelerators

Name	Particle	Energy
Stanford Linear Accelerator (SLAC)	electron	30 GeV 50 GeV (1984)
Fermilab Accelerator (near Chicago)	proton	400 GeV 1000 GeV (1983)
Brookhaven Alternating Gradient Synchrotron	proton	33 GeV
ISABELLE (planned for Brookhaven if funding is made available)	proton colliding beam	400 GeV for each beam (late 1980's)

PARTICULATES Minute bits of solid material suspended in the atmosphere, including those from combustion processes such as ash from fossil fuel and other industrial processes and also those from natural processes such as wind-raised dust, which get incorporated in the air. Also see *Pollution*.

PASCAL A derived SI unit, Pa, for pressure in newtons per square meter. One lb/in^2 equals 6895 Pa, 6.9 kPa. Also see *Conversion Factors*.

PASSENGER EFFICIENCY A measure of the number of people-distance (passenger-miles) which people can be moved for a given quantity of fuel. Representative passenger efficiencies are:

	Passenger-miles per gal	Passenger-km per liter
Diesel train	240	102
Intercity bus	120	51
Vans	100	42
Boeing 747	50	21.3
Compact auto	50	21.3
Standard auto	49	20.8
Luxury auto	35	14.9

For city driving the efficiency of automobiles is half or less the above values for intercity driving. Also see *Freight Efficiency*.

PASSIVE SOLAR DESIGN Those methods of using solar energy that do not use mechanical power to transfer thermal energy units through, and out of a building. The five passive solar systems are direct gain, convective loops, thermal storage walls, roof ponds, and attached greenhouses (Figure P-1).

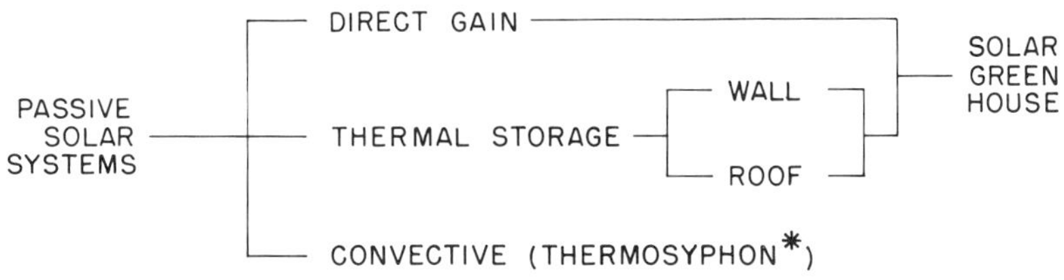

Fig. P-1 Passive solar systems.

PASSIVE SOLAR ENERGY Solar energy utilized without pumps to transfer heat to the point of use. The usual passive solar home designs admit the sun's rays directly into the living space where the energy is circulated by natural convection. A feature often incorporated into passive solar homes is a sizable amount of masonry placed to catch the direct solar radiation, store and release it later when the source of energy from the sun is reduced by clouds or at night. The effect is to even the temperature variations in the living space. The definition of passive solar energy is often modified so as to include systems with fans which help circulate air. Passive solar designs often include fixed or movable shading to prevent over-heating and insulating curtains or shutters to retain heat at night. Also see *Solar Energy*.

Pb Lead.

PBAA Polybutadiene-acrylic acid polymer.

PBN Peroxybutyl nitrate (4-carbon).

pcd Pounds per capita per day, waste.

PCRV Prestressed concrete reactor vessel.

PCS Primary coolant system of a reactor.

PDU Process development unit.

PEAK LOAD The highest electrical or natural gas power demand that must be supplied to the customers in a service area during a period of time such as a day (daily peak) or a year (annual peak). The peak demand is usually an average over a short period of time such as 30 minutes or an hour or, in the case of natural gas, a day. However, seasonal peak refers to the variation of demand with the season in which a longer averaging period is used.

PEAK LOAD SHARING An action taken to reduce the peak load. Also see *Peak Load.*

PEAT A semicarbonized material with an empirical formula of $C_{62}H_{72}O_{24}$, formed in swamps from accumulated plant organic matter. Peat is a precursor of coal. In its natural state peat has a high moisture content, 75% or higher. There are 12,570,000 tons of peat reserve in the U.S.A., principally in Michigan, Minnesota, North Carolina, California, and Florida. (Table P-2.)

TABLE P-2
World Peat Production, 1000 Tonnes, 1979

Country	Agricultural use	Fuel use	Total
USSR	129,500	58,930	188,430
W. Germany	2,000	225	2,225
U.S.A.	785	-	785
Canada	430	-	430
Finland	200	1,840	2,040
Ireland	81	4,615	4,696
Other	1,420	35	1,455
	134,416	65,645	200,061

PELTON WHEEL TURBINE An impulse hydraulic turbine wheel, in which stationary nozzles direct water onto buckets around the periphery of the wheel. With an impulse turbine the pressure head is changed to velocity which impacts and turns the wheel. Pelton turbines are used for heads from 150 to 1675 m (500 to 5500 ft). Also see *Francis; Hydraulic Turbine; Kaplan.*

PERM, PERMEANCE See Water Vapor Transmission Rating.

PERMEABILITY The ease with which fluid moves through porous media, such as gas or oil through sedimentation, gravel and porous rock.

PETRANOL A blend of 10% ethanol and petrol, introduced as a fuel in Australia.

PETROLEUM Also called oil, and that immediately from the earth is called crude oil, or crude petroleum. Petroleum is a liquid formed from vegetable and animal matter in the earth, found in sedimentary rocks and sand and formed as a result of heat and pressure applied for a million years or more. Petroleum is a mixture of hydrocarbons, primarily alkanes, from which many products are made by fractional distillation: Gases, gasoline, fuel oil, and asphalt. There is considerable variation in the content of crude oil, but generally in the range of 82-87% carbon, 10-14% hydrogen by weight plus other minor elements.

The world consumption of petroleum in 1976 was 2900 million tonnes (21.5 x 10^9 barrels). Also see *Alkanes; Distillates; Fractional Distillation; Paraffin Series.*

PFB, PFBC Pressurized fluidized bed; pressurized fluidized bed combustion.

PGC Pyrolysis-gasification-combustion (turbine system).

pH A measure of the concentration of hydrogen ions in a solution. Quantitatively

$$pH = \log_{10} \frac{1}{[H^+]}$$

where $C_H{}^+$ is the number of hydrogen ions per cm^3 of solution. Values of pH below 7 indicate acid and values above 7, base.

PHASE MEDIA CHANGE (PMC) See Latent Heat Storage.

PHOTOCATALYSIS Chemical change brought about by exposure of a substance to light in the presence of a catalyst in which the radiated material is decomposed.

PHOTON A quantum of electromagnetic radiation which has energy equal to the product of the frequency of vibration and Planck's constant. Photons are produced during collisions between electrons and nuclei and deexcitation of nuclei, molecules, and atoms. Also see *Planck's Constant; Quantum.*

PHOTOSATURATION The light intensity beyond which there is no increase in photosynthesis for a plant.

PHOTOSYNTHESIS The formation of carbohydrates in a three-step oxidation-reduction reaction of carbon dioxide and water in the presence of solar energy and catalysts in green plants. The three steps or reactions are: Evolution of oxygen from water, transfer of hydrogen atoms, and reduction of carbon dioxide to carbohydrates. The photosynthetically active radiation region is between 400 and 700 nm (or mμ) wavelength comprising 43% of the energy in the electromagnetic spectrum. Also see *CAM; C_4.*

PHOTOTHERMAL A system for converting solar energy into heat, such as solar collector, solar pond, and solar still. Also see *Flat-Plate Collector; Focusing Collector; Solar Heating; Solar Pond.*

PHOTOVOLTAIC Referring to a process by which light is converted into electricity. In a photovoltaic cell, light striking the active surface of the cell is absorbed resulting in development of a voltage across different parts of the cell. If these parts are connected together through an external electrical circuit, current flows between them and energy is transferred to the external load. These are also called solar cells. Materials most commonly touted for photovoltaic solar cells are silicon, gallium arsenide, and cadmium sulfide. Conversion efficiencies are usually in the range of 5-15%.

PHS Public Health Service (USDHEW).

PHWR Pressurized heavy-water-moderated and cooled reaction.

PILE REACTION The characteristic nuclear chain reaction occurring in a pile (nuclear reactor). Figure P-2 illustrates this reaction including the side effects that lead to Pu-239 formation. Also see *Chain Reactions*.

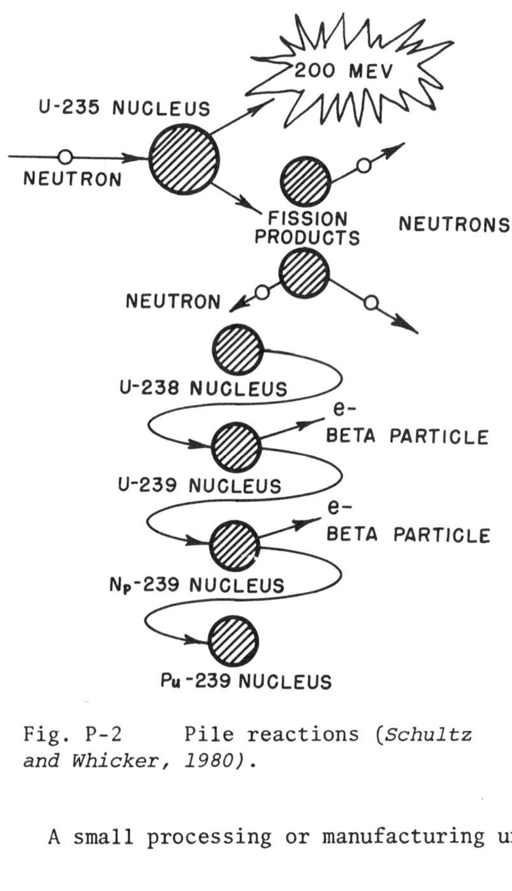

Fig. P-2 Pile reactions (*Schultz and Whicker, 1980*).

PILOT PLANT A small processing or manufacturing unit, larger than laboratory apparatus and smaller than an operating plant. The design of a pilot plant is based on the best information available from bench, laboratory and research results. Information developed by operating a pilot plant is used to evaluate the feasibility and provide the basic design information for full scale operation. Also see *Scale-Up of Equipment*.

PIPELINE A pipe used to transport material from one place to another. Pipelines are widely used in the U.S.A. for transport of gas, liquefied gas, crude oil, and refined products. They are variously placed underground, overland, and in water bodies. Pipelines vary in diameter from a fraction of an inch up to four feet. Large diameter lines generally are used for long distance transport, medium diameters for collecting fuels, and small diameters for delivery to consumers. There are about 175,000 miles of pipeline for oil and one million miles of pipeline for gas in the U.S.A. Energy for pipeline transportation of liquid freight is two-thirds that required for railroads and one-fifth that required for trucks.

PIPELINE QUALITY GAS Relatively pure synthetic gas that can be introduced into a natural gas pipeline without seriously degrading the quality of the gas in the line. Generally the quality required is about 90% methane. Medium-Btu gas, often high in carbon monoxide and hydrogen, may be upgraded by methanation to produce methane. Also see *Methanation*.

PITCH A bottom product resulting from the refining of crude oil and the distillation of coal tar. Pitch tends to have the viscous behavior of a solid in contrast to tar which behaves like a thick liquid. It is sometimes considered to be a grade of asphalt and is a component of tar. Pitch may be obtained from plant material such as the resin from the sap of coniferous trees. It is used as a binder for coal dust when making briquettes. Also see *Asphalt; Bitumen; Tar*.

PITCH LAKE A lake of asphalt covering 45 hectares (113 acres) located on the island of Trinidad. The asphalt or pitch seeps into the lake from a reservoir of material surrounding the area. The lake is liquid near the center, contains pools of water, and is hard near the edges. An odor is emitted particularly as the material is worked in the heat of the day. The lake has been a commercial source of asphalt since the 1700's and in most recent years has yielded 150,000 tons per year.

PITCHBLENDE A heavy mineral which is a high grade uranium ore and often also contains lead, thorium, and rare earth elements. The mineral yields 50-80% uranium, has a high specific gravity and varies in color from greenish-brown to black. The major sources are Canada, Czechoslovakia, United States, South Africa, Zaire, and Australia. Also see *Pitch*.

PLANCK'S CONSTANT A numerical constant usually designated by h.

When it is multiplied by the frequency of electromagnetic radiation, v, the product is the energy contained in a quantum of that type of radiation. The value of h is 6.6255×10^{-34} Js (6.2795×10^{-37} Btu sec). Also see *Quantum*.

PLANT EFFICIENCY (thermal power) The ratio of energy output to energy input for a power plant expressed as a percent, (output ÷ input) x 100% = efficiency in percent. Also see *Efficiency*.

PLASMA A gaseous state of matter consisting of positive and negative ions in approximately equal number. Exposure to plasma radiation may affect the permeability of biological materials, the physical properties of fiber, and the chemical properties of organic matter. Also see *Magnetic Containment; Magnetic Pressure; Magnetic Pumping*.

PLT Princeton large torus.

PLUTONIUM (Pu) A man-made element with atomic number 94. The most common isotope is ^{239}Pu, but ^{240}Pu and ^{241}Pu also are produced in nuclear reactors. The nuclear properties of ^{239}Pu make it a good nuclear fuel and nuclear explosive. ^{239}Pu is radioactive and decays by alpha emission with a half-life of 24,400 years.

Pm Promethium.

PMC Phase media change.

POISE A measure of absolute viscosity in the cgs system of units. One poise is one dyne-second/cm^2. One kg/ms = 10 dyne sec/cm^2 (or poise) = 1 Ns/m^2. The centipoise equal to 0.01 poise is commonly used as a measure of viscosity.

POLLUTANT STANDARD INDEX (PSI) An index of air pollution developed by EPA to represent air quality. A PSI number value of 0 is good; 50, moderate; 100, unhealthful; 200, very unhealthful (alert); 300, hazardous (warning); 400, emergency; and 500, significant harm. The PSI values are determined from the pollutant levels of total suspended particulates (TSP), SO_2, CO, ozone (O_3), and NO_2. Also see *Air Quality Index*.

POLLUTANTS Potentially damaging materials released into the environment as waste products of man-made or natural processes. About 300 million tons of pollutants are discharged from smokestacks each year in the U.S.A.

POLYCYCLIC (POLYNUCLEAR) AROMATIC HYDROCARBON (PAH) Any of a large number of hydrocarbons contained in crude oil, coal, peat, and lignite. The carbon atoms in these hydrocarbons are arranged in a group of hexagons. Examples of PAH compounds, of which there are thousands, are anthracene, biphenyl, dimethyltetralin, dibenzofuran, fluorene, methylindane, naphthalene, pyrene, and xanthene.

POLYCYCLIC ORGANIC MATTER (POM) Compounds which have multiple fused aromatic rings. Included are polycyclic aromatic hydrocarbons (PAH) and heterocycles of oxygen, sulfur, and nitrogen.

POM Polycyclic organic matter.

POROSITY A measure of the extent to which a material contains internal holes or pores. If the pores are connected, the material may be permeable to liquids and gases, i.e., may be able to absorb and transmit them. If the pores are not connected, the material will not be permeable unless the permeability is associated with atomic scale diffusion mechanisms. The pores of a geological formation may be filled with water or in some cases with oil and gas. The porosity of a formation determines its storage capacity. The technique and amount of removal of the water, oil, or gas depend on the permeability of the formation.

POTENTIAL ENERGY Energy stored in such a way that it can be recovered in a more useful form at a later time. Examples of potential energy are the gravitational potential of water stored in a reservoir, the kinetic energy in a flywheel, and the energy of a compressed gas. Also see *Energy; Kinetic Energy*.

POUND A unit of measure of mass or weight, abbreviated lb. The most common is the avoirdupois pound of 16 avoirdupois ounces or 7000 grains or 0.45359 kg as contrasted to the Troy or apothecaries pound of 12 Troy ounces or 5760 grains or 0.3732 kg.

POWER ADMINISTRATIONS The federal government has established the regional power administrations to market electric energy generated by federal projects in those areas. The five power administrations are the Alaska Power Administration (APA), PO Box 50, Juneau, AK 99801; Bonneville Power Administration (BPA), Portland, OR 97208; Southeastern Power Administration (SEPA), Elberton, GA 30635; Southwestern Power Administration (SPA), Tulsa, OK 74101; and the Western Area Power Administration (WAPA), PO Box 25007, Denver, CO 80221. The Power Administrations are currently part of the Department of Energy.

POWER FACTOR A quantity used to describe the operation of an electric alternating current power system. The power factor is the ratio of the true power, W, dissipated by the system, divided by the apparent power defined as the product of the current, I, and the voltage, E, across it. Thus the power factor is W/EI. A power factor less than one results when the current and voltage are out of phase with each other. Using Θ as the phase angle, cos Θ is also the power factor. A low power factor usually results from coils and windings in the circuit. An additional charge is often made by a utility for power delivered to a circuit with a power factor less than 0.8. A low power factor for an appliance should not be confused with low energy efficiency, defined as energy output divided by energy input. Rather low power factor implies that serving the circuit makes high demands on the capacity of the power generating and distribution system.

POWER GAS   A low-Btu gas, 5.6-9.3 MJ/m^3 (150-250 Btu/ft^3), also called producer gas, usually produced from coal. Power gas is much lower in heat content than coal gas, 16.8-18.7 MJ/m^3 (450-500 Btu/lb). Power gas is used to power engines or for heating. Also see *Coal Gas; Low-Btu Gas; Producer Gas.*

POWER PLANT EFFICIENCY   See Efficiency.

POWER TOWER   A facility for collecting solar energy in which the sun's rays are reflected by mirrors (heliostats) onto a central tower, where water is vaporized into steam and used to drive turbine-generators to produce electricity.

POWER WHEELING   A procedure or practice in the power supply industry in which electric power (energy) belonging to one utility is transferred over transmission lines belonging to another utility. Also see *Grid.*

PPB   Represents parts per billion. The weight or volume of a substance per one billion units of weight or volume of the solution. Equivalent to 1 milligram/m^3 (approximately) at the same weight, or 1 unit in 10^9 units. Also see *PPM.*

PPBR   Program planning budgeting review.

PPH   Pounds per hour.

PPM   Represents parts per million. The weight or volume of a substance per one million units of weight or volume of the solution. 10,000 ppm = 1%. Equivalent to 1 mgm/kg or 1 unit in 10^6 units. Also see *PPB.*

PPN   Peroxypropionyl nitrate (3-carbons).

PRECAMBRIAN   See Era.

PRECISION   Refers to the quality of measurement as stated. The degree to which a measurement represents the true value of the quantity measured. Also see *Accuracy.*

PREFIXES   See Number Prefixes.

PRESSURIZED HEAVY WATER REACTOR   A nuclear reactor in which heavy water under pressure is used as the heat transfer medium. There are several reactors of this type outside the U.S.A., but none in the U.S.A. Heavy water (D_2O) is excellent as a moderator, Figure P-3. Also see *Heavy Water; Moderator.*

PRESSURIZED WATER REACTOR (PWR)   A nuclear reactor in which water under pressure is circulated through the core, then through a heat exchanger to generate steam (Figure P-4). Pressures up to 2000 psia are used to maintain the liquid state. The PWR requires enriched fuel when water is used as the moderator.

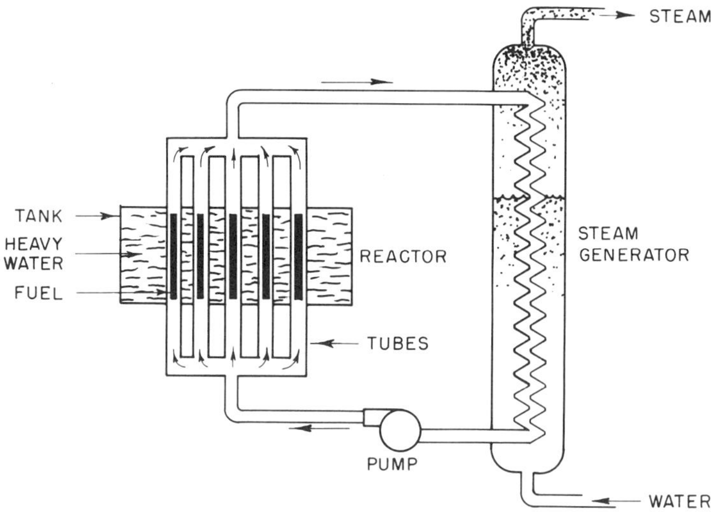

Fig. P-3 Pressurized heavy water reactor.

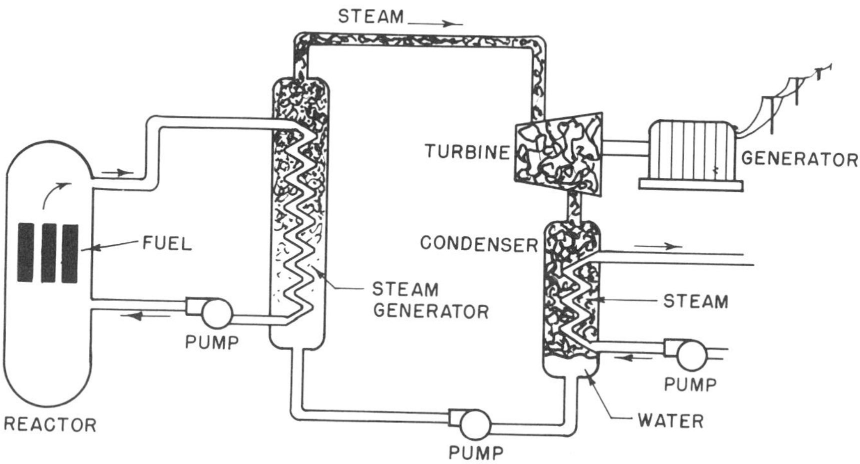

Fig. P-4 Pressurized water reactor (PWR).

PRICE-ANDERSON ACT An act of the Congress of the United States, passed in 1957, which provides for liability insurance and a maximum liability claim ($560 million) in case of a major nuclear accident. The act is under review and may be modified. Liability insurance up to the maximum available is purchased by the industry from private sources and the government provides the remainder up to $560 million. Also see *Nuclear Insurance.*

PRIMARY ENERGY Energy considered to be coming from natural and original sources. Although solar and nuclear energies are considered the most basic energy sources, generally accepted primary sources of energy are solar, coal, oil, natural gas, falling and flowing water, wind, uranium, geothermal, and biomass. Some sources of primary energy are used directly for fuel and some are converted to a secondary form. Examples of the latter are electricity and hydrogen. Also see *Secondary Fuels.*

PRIMARY ENERGY EQUIVALENT The amount of primary energy in some specific form such as coal required to generate a given amount of electricity. This concept is used for calculating and expressing the energy required to generate electricity by different means on an equivalent basis. As an example, the primary energy required to produce electricity at a hydroelectric project can be replaced by the amount of fuel energy required if the electricity were produced in a fossil fuel power plant using 10,000 Btu to produce 1 kWh. Another procedure is to calculate primary energy assuming a conversion efficiency of 33% for converting from a primary fuel to electricity regardless of the actual efficiency involved. Also see *Gross Energy Requirements.*

PRIMARY FUEL A primary energy source used directly, without conversion, as a fuel. The primary fuels are most often burned and used in heating systems or in engines. The major primary fuels are coal, wood, oil, and natural gas. Also see *Primary Energy.*

PRIMARY RECOVERY The first production phase of an oil well. No stimulation of flow is used other than pumping. Also see *Secondary Recovery; Tertiary Recovery.*

PROBO Product-oil-bulk-ore.

PROCESS EFFICIENCY A measure of the efficiency with which some resource is utilized in a process. The resource may be energy, one of the raw materials used in the process, or something else.

PROCESS ENERGY REQUIREMENT The energy that must be supplied to carry out a process.

PRODUCER GAS A low-grade fuel gas 4.6, MJ/m^3 (125 Btu/ft^3), produced by passing air through red hot fuel. The fuel can be coke, coal, sawdust, etc. The carbon in the fuel reacts with the oxygen in the air to produce CO_2 which is then reduced to CO, the principal fuel component of producer gas. If the fuel is wet, the water gas reactions may lead to the presence of hydrogen in the producer gas. The heating value of producer gas is greatly limited by the high concentration (~65%) of inert nitrogen that it contains. The characteristics of producer gas are: Specific gravity relative to air, 0.86; nitrogen, 50%, carbon monoxide, 23%; hydrogen, 17.7%; carbon dioxide, 8.0%. Producer gas is used primarily for heating applications although it can be used for gas engine operation. The term power gas or air gas (England) is sometimes applied to producer gas. Also see *Blast Furnace Gas; Carbon Dioxide; Carbon Monoxide; Hydrogen; Low-Btu Gas; Nitrogen; Power Gas*.

PRODUCT EFFICIENCY The fraction of the inputs used in manufacture that appear in the product. Also used as the heat value of the fuel produced at the output as a percent of the heat value of the material from which the fuel is made, without giving credit to the other products.

PROPANE C_3H_8, a compound of carbon and hydrogen, a colorless gas at normal temperature, used for a fuel for heating and engines. Propane is the principal component, along with butane, of liquefied petroleum gas (LPG). Propane is stored and transported in pressurized vessels as a liquid. Propane has a calorific value of 2400-2600 Btu/ft^3. Physical characteristics are: 4.23 lb/gal; boiling point, -44.5°C (-44°F); flash point, -104°C (-156°F); S.G., 0.509. Also see *Liquefied Petroleum Gas (LPG)*.

PROPANOL The alcohols which have the molecular formula of C_3H_8OH, known as propyl alcohol or 1-propanol (n-Propanol). Made from ethylene or propane, and obtained as a byproduct of distillation of fermented products.

PROPPANT or PROPPING AGENT A granular material used to facilitate production of gas from a reservoir by maintaining the openings through which the gas is removed—sand, glass balls, walnut shells, sintered bauxite is used. The propping agent is injected into the reservoir by a high pressure fluid to fracture and enlarge the reservoir openings. Also see *Fracturing*.

PROTEINS The complex chemical nitrogenous compounds of plants and animals which are necessary for food for humans and animals. When heated in the presence of water, denaturization occurs, followed by coagulation, flocculation, and precipitation with continued heating. Enzymes are proteins, which are inactivated with appropriate heating. There are at least 80 amino acids of which 20 are building blocks for proteins and eight are essential for human diet.

PROTON A positively charged nuclear particle. The nucleus of an ordinary hydrogen atom is a proton. Protons also are constituents of all other atomic nuclei. The number of protons in the nucleus of an atom is called the atomic number and identifies the atom involved. The number of electrons in an atom is equal to the number of protons. The proton and electron have charges equal in size but opposite in sign. The proton mass is 1837 times the electron mass, 1.7×10^{-27} kg. Also see *Electron; MeV*.

PROXIMATE ANALYSIS The chemical analysis of coal, and other fuels and minerals, in which the percentages of certain elements or constituents are determined. The determinations are accurately made, so proximate does not mean approximate. The elements or constituents to be included and the procedures to be used to determine them are standardized by an agency, association, technical society or group, such as ASTM. For coal the proximate analysis usually includes fixed carbon, moisture, volatile matter, and ash. In contrast, ultimate analysis gives more exact and complete information on the constituents.

PSI See Pollutant Standard Index.

PSI, PSIA, PSIG Psi (pounds per square inch): Measure of pressure. Psia (pounds per square inch absolute): Total pressure obtained by adding atmospheric pressure to psig. Psig (pounds per square inch gage) is the pressure that a normal pressure gage measures. The difference between the pressure in the vessel, to which the gage is connected and the atmospheric pressure is the psig. When psi alone is used, it usually refers to psig.

PSPE Photosynthetic production enhancement.

Pu Plutonium.

PUD Public Utility District.

PUMPED STORAGE An energy storage concept. A pumped storage system consists of two large water reservoirs at different elevations. They are joined by a water conduit containing a reversible electric pump and electric generator unit. The system generates electric power when water flows from the upper reservoir through the conduit and generator to the lower reservoir. Energy is stored as potential energy when the electric pump transfers water from the lower to the upper reservoir (Figure P-5). Also see *Hydroelectric*.

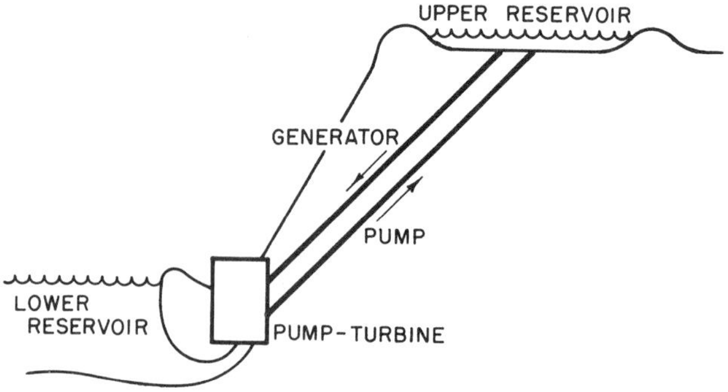

Fig. P-5 Pumped storage.

PV, pv Photovoltaic.

PVDC Polyvinylidene chloride.

PVS Photovoltaic system.

PWR Pressurized water reactor.

PYRITE The principal form of iron sulfide, FeS_2, found in coal.

PYROLYSIS The decomposition by heat of the organic fraction of a liquid or solid, usually at 750°F or higher, in an oxygen-deficient atmosphere. Pyrolytic processes produce gases and char. Pyrolysis is used in producing gases and liquids from coal, shale, and solid organic wastes. Also see *COED; Garrett; Synthetic Crude Oil.*

PYROMETER An instrument that measures high temperatures. There are several kinds of pyrometers. The optical pyrometer measures the temperatures of very hot objects by observing color. As the temperature of a hot, glowing object increases, the color of the radiation it emits, shifts from red toward white. Other pyrometers use thermoelectric voltages or electrical resistance to measure temperature.

Q

Q   A letter used to represent a unit of energy equal to a quintillion (10^{18}) Btu. In terms of energy, the Q was originally defined as the energy equivalent to 30 x 10^9 tons of charcoal. The total annual U.S.A. energy consumption is less than 0.1 Q.

$$Q = 10^{18} \text{ Btu}$$
$$= 1.05 \times 10^{21} \text{ joule}$$

Recently several authors have used Q for 10^{15} Btu instead of q or mQ used by other authors. Q is also used for representing the charge of a capacitor, water flow, magnification factor, and total heat transfer. Also see *Quint*.

QUAD   A unit of energy equal to a quadrillion (10^{15}) Btu. Also see *Quadrillion; Quintillion*.

QUADRILLION   A number used to represent large values, particularly for energy. A quadrillion, quad, is used for 10^{15} units in the U.S.A. and 10^{24} in Great Britain. Some authors identify 10^{15} as 1 mill Q or mQ or q. One quad is represented by

0.5 million barrels of oil per day for a year
1 trillion (10^{12}) cubic feet of natural gas
40 million tonnes of coal
8 billion gallons of gasoline

Also see *Q; Quintillion*.

QUALITY FACTOR   A factor by which an absorbed dose of uniformly distributed radiation should be multiplied to obtain an equivalent dose that determines the biological effect of the radiation. The quality factor for heavy ionizing particles is directly related to the linear energy transfer of the radiation. The quality factor for x-rays and electrons is 1.

QUANTUM A discrete amount of energy absorbed, emitted, or possessed by an atom, nucleus, molecule, or other unit of matter. The energy, E, in a quantum of electromagnetic radiation is $E = h\nu$, in which h is Planck's constant, 6.6255×10^{-34} Js (6.2795×10^{-37} Btu/sec), and ν is the frequency of radiation. Also see *Photon; Planck's Constant.*

QUEBRACKS; QUEBRACHO A thinning agent of tannic acid ($C_{14}H_{10}O_9$) added to a well drilling fluid to provide favorable physical properties, principally viscosity, for drilling.

QUINT A quintillion (10^{18}) Btu. Also called a Q. Also see *Q.*

QUINTILLION A number, represented by Q, quint, for 10^{18} units in the U.S.A. and 10^{30} units in Great Britain. Used to represent large quantities, such as Btu of energy, world inventories, etc. Also see *Q; Quadrillion.*

Q-VALUE (nuclear) The total kinetic energy of the products of a nuclear reaction minus the total kinetic energy of the reactants. The Q-value measures the energy released by the nuclear reaction.

R

RAD A unit of radiation energy dose absorbed in tissue or other material. One rad is an absorbed dose of 100 ergs per gram of tissue or other matter, or 0.01 kJ/g, or 0.01 Grey (Gy). Also see *Rep*.

RADIANT HEATING See Infrared Heating.

RADIATION Energy transmitted through a medium by waves or particles. Thermal, solar, light, radio, microwave, x-ray, and gamma radiations are all transmitted by electromagnetic waves. Sound radiation is transmitted by pressure waves in the air or other sound transmitting medium. Streams of nuclear particles such as alpha particles and neutrons are examples of nuclear radiations that are not waves (except in the quantum mechanical sense). Also see *Electromagnetic; Infrared; Irradiation; Low Level Radiation; Transuranic Wastes*.

RADIATION, NUCLEAR Any of several kinds of particles and rays emitted by nuclei. Alpha and beta particles and gamma rays are the most common types of nuclear radiations. Also see *Radiation*.

RADIATION EXPOSURE Estimates of radiation exposure are listed in Table R-1. These exposures are not directly comparable as some radiation is directed to a part of the body and others to the entire body. Also see *Risks of Different Energy Sources*.

RADICAL In chemistry, radical refers to a group of atoms which maintains identity in a chemical reaction. A radical is distinguished from an ion. An ion has a charge while a radical may or not have a charge. Examples are: Ammonia radical, NH_4; water radical, H_2O; and the OH group which may be a radical, -OH, or an ion, OH^-.

RADIOACTIVE DECAY A process during which a nucleus spontaneously changes to a different nucleus with an accompanying emission of a particle or particles. Each radioactive decay has its own

TABLE R-1
Radiation Exposures to the Body, millirems

Flight from Los Angeles to Paris, cosmic	4.8
Chest X-ray	22
Dental X-ray	910
Breast mammography	1,500
Occupational exposure limit, 1 yr	5,000
Barium enema	8,000
Radiation treatment for bone cancer	6,000,000

characteristic rate of occurrence, which can be specified by its half-life and characteristic nuclear radiations.

RADIOACTIVE NUCLIDE An atomic species which has a specific type of radioactive nucleus. An example of a radionuclide is tritium, which is a radioactive form of hydrogen. The nucleus of a tritium atom contains two neutrons and one proton. Also see *Nuclide; Radionuclide*.

RADIOCARBON DATING A technique used to determine the age of ancient biological materials. The procedure depends upon the fact that living biological material contains carbon that comes from the atmosphere either directly from carbon dioxide assimilated by plants or indirectly from plant material eaten by animals. Atmospheric carbon contains a small proportion of radioactive ^{14}C, which is produced in the air by cosmic rays. Living material thus contains the same proportion of ^{14}C as the atmosphere. However, when an organism dies, the ingestion of new carbon ceases and the proportion of ^{14}C declines as the radioactive decay proceeds. The time after death of material is determined by taking a small sample, converting the carbon to carbon dioxide, and measuring the proportion of the carbon that is ^{14}C. The half-life of ^{14}C is 5730 years so that the technique is most suited to determining the age of materials several thousand years old.

RADIOISOTOPE A radioactive nuclide. Two isotopes differ in atomic mass but not atomic number. One may be radioactive while the other is not. Radioisotopes may occur naturally, for example, potassium-40, thorium-232, uranium-235, and uranium-238, or may be artificially produced by bombardment with high-energy particles, for example, carbon-14 (^{14}C), which is widely used as a tracer. Also see *Isotope; Nuclide*.

RADIONUCLIDE A nuclide which is radioactive. Also used loosely or generally as radioisotope. An atomic species with a radioactive nucleus. Also see *Nuclide; Radioactive Nuclide*.

RADIOSITY The total radiant energy leaving a surface in terms of energy per unit area of surface, which includes reflected as well as emitted energy. The total radiosity of a surface includes all radiation leaving a surface, regardless of direction of radiation or spectrum involved.

RADIUM (Ra) A naturally occurring radioactive element with atomic number 88. The naturally occurring isotope, ^{226}Ra, has a half-life of 1600 y and decays to an isotope of radon gas, ^{222}Rn. Also see *Half-Life*.

RADON (Rn) A naturally occurring radioactive element with atomic number 86. The naturally occurring isotope, ^{222}Rn, has a half-life of 3.8 days and decays by alpha emission. Radon is a gas that does not form chemical compounds and tends to diffuse out of the material in which it is formed. Also see *Half-Life*.

RANKINE SCALE A temperature scale with zero point at absolute zero and degree interval equal to the Fahrenheit scale interval, represented by °R. The absolute zero on the Rankine scale corresponds to -459.6°F.

RARE Roadless Area Review and Evaluation.

RARE EARTH ELEMENTS Those elements which have an atomic number between 57 and 71, known as lanthanides, and between 89 and 103, known as actinides. The actinides are radioactive rare earth elements. Also see *Actinides*.

RAW GAS Natural gas as it exists or is produced from a gas field. Also see *Natural Gas*.

RAW SHALE Oil shale before it has been retorted.

RBE Relative biological effectiveness.

RD&D Research, development and demonstration.

RDF Refuse-derived fuel.

RDT&E Research, development, testing and evaluation.

REA Rural Electrification Administration.

REACTION TURBINE See Dariaz Turbine.

RECOVERY FACTOR The portion of oil stated in percentage or decimal, which can be removed from an oil well using primary (natural), secondary (adding fluid), and tertiary (solvent) methods. The recovery factor may vary from 10 to 80%. Also see *Oil Recovery; Primary Recovery; Secondary Recovery; Tertiary Recovery*.

REFINERY GASES Those gases produced during cracking of petroleum which come off at a low temperature of 0°C (32°F) and below, which are principally methane, ethane, propane, and butane, as well as hydrogen and carbon dioxide. Propane and butane are liquefied and sold as liquefied petroleum gas (LPG). Also see *Off-Gas*.

REFINING   As applied to crude petroleum, the treatment such as by distillation and cracking to provide various components or fractions, such as upper, medium, and lower distillates, which can be treated further by reforming for specific commercial products, such as gasoline, diesel, kerosene, lubricants, etc. Also see *Distillates; Reforming*.

REFLECTION   The change in direction of electromagnetic radiation when it strikes a surface. Reflection may be regular or diffuse as shown in Figure R-1. The extent of the reflection in some situations is given in Table R-2.

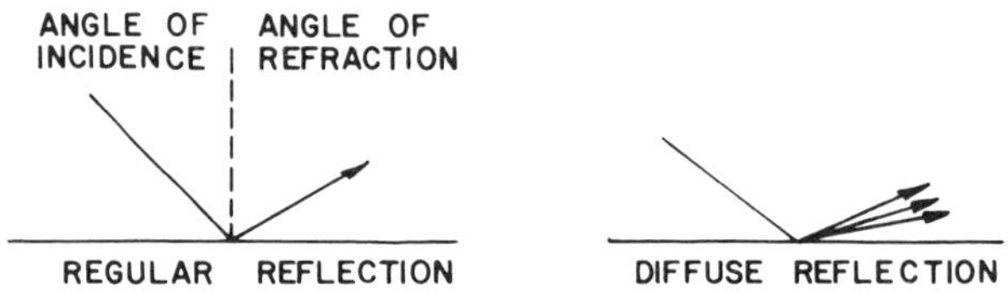

Fig. R-1   Reflection.

TABLE R-2
Reflection of Incandescent Light

Snow	0.93
Steel	0.55
Black paper	0.05-0.06
Aluminum	0.77-0.81

REFORMING   A secondary refining process to change the initial product so as to (1) improve the ignition properties, (2) change the proportion of constituents, (3) introduce an element or compound that is missing, and/or (4) prepare a particular chemical feedstock. Catalytic reforming (using a catalyst, usually metal) and thermal reforming (temperature and pressure) are used. Also see *Catalytic Reforming; Refining; Thermal Reforming*.

REFRACTION   The change in direction of an electromagnetic wave as its energy passes from one material to another or as it moves through a material whose properties (such as density) vary from place to place (Figure R-2).

REGENERATION   A procedure to re-use the thermal energy in a process

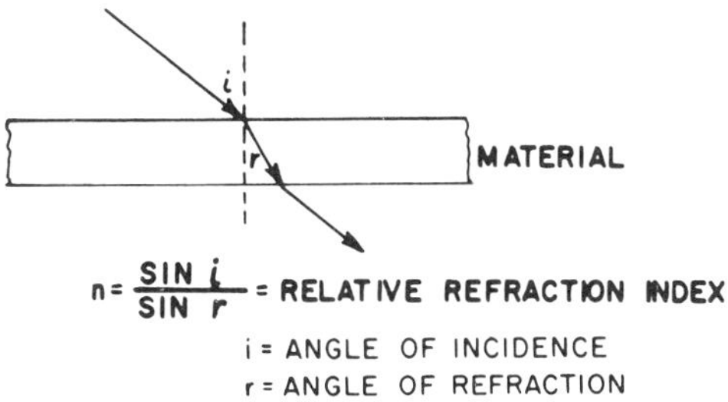

Fig. R-2 Refraction.

to increase the thermal efficiency and economy. Regeneration usually requires additional equipment, principally heat exchangers. The most common example of regeneration is a transfer of some of the heat from an outgoing hot processed product to increase the temperature of the material entering the process. A regeneration efficiency (thermal) of 50 to 90% may be economically obtained depending on the temperature, product, cost of energy, and cost of purchasing, maintaining, and cleaning the additional heat exchangers. In the Rankine cycle, steam may be removed at various locations to heat incoming feed water in what is called a regenerative Rankine Cycle.

REID VAPOR PRESSURE (RVP) The approximate vapor pressure of fuel at 37.8°C (100°F) used to describe some jet fuels. JP-3, the commercial jet fuel which originally was most widely used, has an RVP of 6; JP-4, 2½; aviation gasoline, 6.

RELATIVE BIOLOGICAL EFFECTIVENESS, RBE (radiobiology) The ratio of the dose required to produce a given effect with x-rays to the dose required to produce the same effect with the radiation under consideration. The relative biological effectiveness depends on the type and energy of the radiation under consideration and on the type and extent of the biological effect involved.

REM Roentgen equivalent man.

RENEWABLE ENERGY RESOURCES Those sources of energy which do not rely on the mineral resources of limited quantity: Coal, oil, gas, and uranium. Renewable sources depend on the energy from the sun or gravitational forces. The renewable energy resources are generally considered to include: Solar, biomass, wind, wave, tidal, and hydro. Coal, natural gas, and oil are examples of resources that are not renewable. Also see *Alternative Energy Sources; Ambient Energy Sources*.

REP See Roentgen equivalent physical.

REPROCESSING The process of removing unused fuel from spent (used) fuel elements and incorporating the fuel into new elements.

RESEARCH OCTANE NUMBER See Octane Number.

RESERVES A term usually used to indicate the amount of fossil fuel remaining in the earth. Examples are measured reserves; initial reserves; proven reserves; indicated reserves; inferred reserves; deferred reserves; demonstrated reserves; ultimate reserves; probable reserves; possible reserves; and established reserves, which includes proven reserves plus some probable reserves. Other terms used include remaining reserves, in-place reserves, recoverable reserves, and marketable reserves. The term is also used to describe biomass energy sources.

RESIDUAL OIL See Residuum.

RESIDUUM The material remaining as heavy liquid or solid following distillation or cracking processes. The residuum from crude oil has a boiling point of at least 469°C (875°F), a molecular weight of 492 to 1262, and molecules with 36 to 90 carbon atoms. Used for heating or producing steam to make electricity. Also known as residual oil.

RESPIRATION The process of gas exchange in living organisms in which oxygen is moved into the system and carbon dioxide moved out of the system. Carbon dioxide is produced through slow utilization of carbon in a metabolic process by an organism. Also see *Photosynthesis*.

RETROFIT Changing or modifying a device to accomplish a new objective, such as retrofitting a conventional building to utilize solar energy for heating.

RFF Resources for the Future, an organization located in Washington, DC.

RHEOLOGY A study of the deformation and flow of materials. Measurement of rheological properties may provide an indirect measure of a variable related to processing, wetting, drying, and moisture content of a material.

RISKS OF DIFFERENT ENERGY SOURCES The comparative risks from different activities or exposures which incur a one-in-a-million risk of death are:

- 600 km travel by air
- 100 km travel by automobile
- 3/4 cigarette
- 1/3 hour of life at age 60
- exposure to radiation by living in vicinity of a nuclear plant for three years

Source: IAE Bull. 22, 1980.

Also see *Radiation Exposure*.

ROCKY MOUNTAIN OVERTHRUST BELT See Overthrust Belts.

ROENTGEN The exposure to x- or gamma radiation that will produce 2.58×10^{-4} coulombs of ions of each sign per kilogram of air. An older and nearly equivalent definition is that the roentgen is the exposure to x- or gamma radiation that will produce one electrostatic unit of ions of each sign per cubic centimeter of dry air at standard temperature (0°C) and pressure (one atmosphere). The roentgen is a measure of exposure (ionization in air) rather than dose (energy absorbed). Also see *Dose; Exposure; Rad; Rem.*

ROENTGEN EQUIVALENT MAN (Rem) The unit of dose equivalent for radiation. The dose equivalent in rems is numerically equal to the absorbed dose in rads multiplied by the quality factor, the dose distribution factor, and any other necessary modifying factors. The rem is designed to be an indicator of the biological effect of radiation. Also see *Absorbed Dose; Dose Equivalent; Quality Factor; Rad; Radiation Exposure; Rep.*

ROENTGEN EQUIVALENT PHYSICAL (Rep) That amount of any type of nuclear radiation from which soft tissue will absorb an energy of 93 ergs per gram. This unit is now obsolete and has been replaced by the rad.

RON Research octane number. See Octane Number.

ROSE Residuum Oil Supercritical Extraction.

RUBBLE The product resulting from the breaking up of consolidated materials usually with explosives.

RUN-OF-THE-MINE A term used to describe coal, shale, or ore as it comes from the mine before it is sized or otherwise processed.

R-VALUE A number used to represent the resistance to transfer of heat, or insulation value, of materials or combinations of materials used for construction—buildings, pipelines, and storages. The R-value of a material is most often expressed in English units

$$\frac{ft^2 hr\ °F}{Btu}$$

The R-value of materials can be added to obtain the R-value of a wall. The R-value of an inch of rock wool, mineral wool, glass wool, and similar insulation ranges between 3 and 4; thus six inches would have an R-value of 18 to 24. The larger the R-value the greater the resistance to heat transfer and the less heat lost or gained. Construction codes in some areas now specify a minimum R-value for walls and ceilings for buildings. In customary units recommended R-values vary from a low of 5 to 10, to as high as 35 for buildings in cold climates heated with electricity and for buildings in hot climates with air conditioning.

The R-value is the sum of the reciprocals of the wall conductivity,

$$\frac{x}{k}$$

and conductance coefficients of standard members,

$$\frac{1}{C}$$

and the film coefficients of air,

$$\frac{1}{f}$$

Thus, an R-value is calculated,

$$R = \frac{1}{f_o} + \frac{x_1}{k_1} + \frac{x_2}{k_h} + \frac{1}{C} + \ldots + \frac{1}{f_i}$$

where f_o = outside film coefficient

x_1, x_2 = thickness of wall components
k_1, k_h = conductivity of wall components
C = conductance of wall components of a given material and thickness

Representative values of f, k, and C for determining R-values

f_o,	outside air film, wind of 15 mph	6.00
f_i,	inside air film, nearly still air	1.65
C,	conductance of air space over 3/4 in	1.10
C,	conductance of double pane window	0.61
C,	conductance of 8 in concrete block	0.90
k,	per inch of thickness of	
	plaster board	1.41
	and flexible types of insulation of mineral wool, rock wool, glass wool	0.27
	insulation board	0.34
	concrete, stone terrazzo	12.5
	wood	1.0

In customary units R-values are in hr ft^2 °F/Btu; in SI units, m^{2}K/W. 1 hr ft^2 °F/Btu = 0.176 m^{2}K/W. A closely related quantity is U = 1/R. Also see *Overall Heat Transfer Coefficient*.

RVP Reid vapor pressure.

S

SAE Society of Automotive Engineers, 400 Commonwealth Drive, Warrendale, PA 15096.

SANDS A general term used to identify porous rock and sand which holds hydrocarbons. Also see *Sandstone*.

SANDSTONE A porous sedimentary rock composed of grains of sand compressed or cemented together. Sandstone deposits are of particular importance to the petroleum industry because most deposits of petroleum are located in sand and sandstone. Also see *Sands*.

SASOL The acronym for the South African Coal, Oil, and Gas Corporation which has developed Sasol I (1955) and II plants for converting coal to liquid fuels. The coal is first gasified using the Lurgi process, then liquefied using the Fischer-Tropsch process. Also see *Coal Liquefaction; Fischer-Tropsch; Lurgi*.

SATURATED HYDROCARBON A hydrocarbon which has an arrangement in which all of the adjacent carbon atoms are connected by one bond or valence. Additional hydrogen cannot be taken on by the carbon without splitting the molecule. Another element cannot be taken on without giving up hydrogen. Saturated hydrocarbons do not have a double or triple bond. The alkanes (paraffins) are saturated hydrocarbons. Also see *Alkanes; Paraffins*.

SATURATION CURVE A graphical representation of the temperature-entropy relationship between a liquid and vapor of that liquid. A saturated vapor has a temperature corresponding to the temperature of boiling of the liquid (Figure S-1).

SAVONIUS ROTOR A hemicylinder rotor for a windmill which rotates around a vertical axis (Figure S-2). An advantage, in contrast to conventional wheels, is that the rotor can accept the wind from any direction without the need for a swivel arrangement. The efficiency

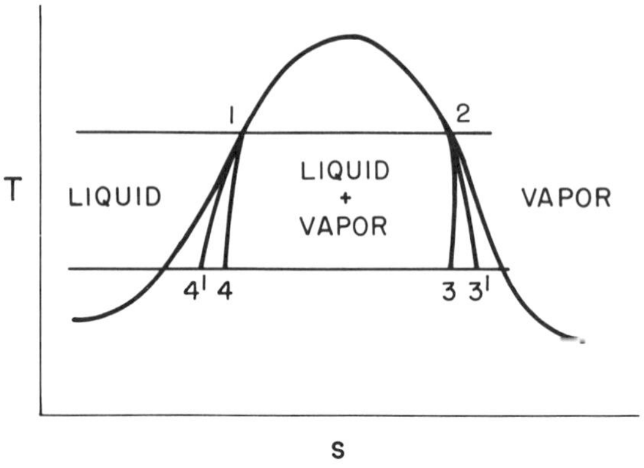

Fig. S-1 Saturation curve.

of the hemicylinder rotor is low, 10 to 15%; as contrasted to 30 to 40% for the multibladed windmill. Also see *Darrieus Rotor; Wind Energy; Wind Turbine; Windmill.*

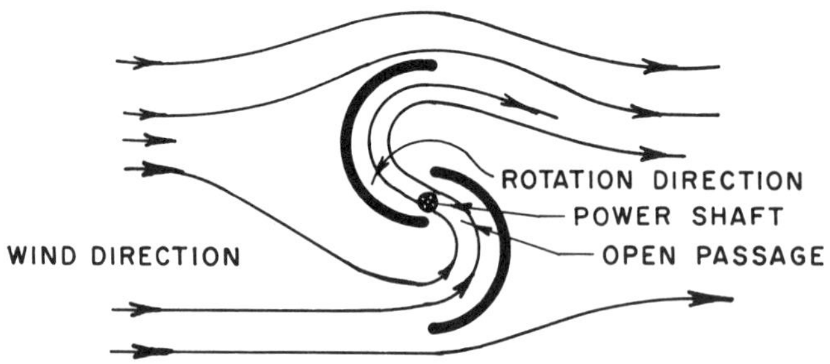

Fig. S-2 Savonius rotor.

SCALE-UP OF EQUIPMENT The general term applied to larger equipment based on experience with smaller units. Typical scale-up involves small oven or retort sample testing, laboratory scale, pilot plant, semi-commercial unit, and full-sized plant. The oven test may be used to evaluate the effect of basic phenomena such as heat or circulation rates on reaction rates. The laboratory model is used to begin work on a very small scale. The pilot plant is generally instrumented sufficiently to permit uniform results enabling market analysis of the product. The pilot plant is the first unit to use equipment and materials similar to those expected to be suitable for the finished plant. Proper pilot plant testing usually allows skipping the semi-works state for design parameters, limiting its use to cases where developing markets preclude major investment at once. Scale-up magnitudes from 1 to 5 have been replaced by values approaching 1 to 100 with increasing knowledge of the fundamental factors affecting processes. Computer simulation has been very helpful in studying the effect of various design parameters. Parameters of major significance in processing include: Pressure, air circulation rate, moisture content of air in and out, temperature of air and material, state of subdivision of the material being processed, effect of depth of bed or layer, pattern of air circulation, catalysts, etc. Also see *Pilot Plant*.

SCFD, scfd Standard cubic feet per day.

SCINTILLATOR A material that converts ionization energy produced within it into light energy. A scintillator can be used as part of a counter of nuclear radiation.

SCOT Shell Claus Off-gas Treating.

SCR Semiconductor (or silicon or selenium) controlled rectifier.

SCS Soil Conservation Service.

SCS Silicon controlled switch.

SDWA Safe Drinking Water Act.

SEA Science and Education Administration (USDA).

SEAM COAL Coal ranked between bituminous and anthracite, 91-93% carbon and 8-22% volatile matter.

SECOND LAW EFFICIENCY The ratio of theoretical minimum amount of energy needed to carry out a task to the actual amount used to carry out the task. The second law efficiency is usually much less than the typical first law efficiency of useful work output divided by work input. Also see *Efficiency*.

SECOND LAW OF THERMODYNAMICS The law which states that it is impossible to construct a device that operates in a cycle and produces no effect other than the transfer of heat from a cooler body to a hotter body. Stated another way, it states that it is impossible to construct a device with will operate in a cycle and produce no effect other than the raising of a weight and the exchange of heat with a single reservoir. This law sets a limit on the percentage of input heat energy which can be converted to useful work energy. Also see *Entropy; First Law*.

SECONDARY FUELS Fuels which have been produced by processing primary energy sources. In many cases the secondary fuels have more desirable properties than the primary sources. Secondary fuels may be solid, such as char or coke; liquid, such as gasoline, diesel or fuel oil; or gas, such as hydrogen or carbon monoxide. Electricity is considered a secondary energy (fuel). Also see *Fuel; Primary Energy.*

SECONDARY OIL RECOVERY The practice of extracting additional oil by high-pressure steam injection, water flooding, gas loading, or air from a well following the use of natural flow in primary recovery. Primary recovery may remove 25-40% of the oil, with an additional 10-25% removed by secondary recovery procedures. Using steam produced by oil may require one bbl of oil to recover three bbl of oil. Further efforts using tertiary recovery procedures may be used to remove an additional amount up to 25% at the most, with present techniques. Also see *Enhanced Oil Recovery; Oil Recovery; Primary Recovery; Tertiary Recovery.*

SEDIMENTARY Referring to deposits formed by the settling of organic and inorganic materials to the bottoms of water bodies. These deposits, often consolidated into rock layers over the course of geologic time, are potential sources of hydrocarbon fuels. Also see *Sand.*

SEIA Solar Energy Industries Association.

SEMICONDUCTOR (or silicon or selenium) CONTROLLED RECTIFIER A thyristor (semiconductor) device for converting alternating current to direct current electricity.

SEMI-WATER GAS See Water Gas.

SENSIBLE HEAT Thermal energy or heat which when added to a substance increases its temperature. Sensible heat added to a substance can be sensed by a thermometer.

SEPA Southeastern Power Administration. See Power Administrations.

SEPARATION See Dialysis; Filter; Liquation.

SERI Solar Energy Research Institute.

SERPY The amount of solar energy received per year, 1.5 x 10^{15} MWh, reaching the outer atmosphere of the earth.

SEVEN SISTERS The name given to the seven large oil companies which refine and market crude oil worldwide: Exxon, Texaco, Mobil, Standard Oil of California, Gulf, British Petroleum, and Royal Dutch Shell.

SGHWR Steam generating heavy water reactor.

SGP Shell gasification plan.

SHAC Solar heating and cooling.

SHALE Rock formed from clay incorporating plants which can be used as a source of oil. Also see *Cannel Shale*.

SHALE OIL The organic oil obtained by pyrolysis of oil shale at 482°C (900°F). Shale oil is 80-86% carbon and has a C/H ratio of 7-9.

SHIFT CONVERSION A process for producing gas with a desired concentration of carbon monoxide from gases derived from gasification. A carbon monoxide enriched gas is saturated with steam and passed through a catalytic reactor where the carbon monoxide reacts with steam and shifts the hydrogen to carbon monoxide ratio.

SHORT CORD See Face Cord or Facecord.

SI UNITS or INTERNATIONAL SYSTEM OF UNITS A standard set of units in which physical quantities can be measured. This set of units was established as the international standard at the 11th General Conference on Weights and Measures in Paris, held in 1960. The basic units are the meter, m; kilogram, kg; second, s; ampere, A; kelvin, K; candela, cd; and mole, mol; and the subdivisions and multiples thereof. Secondary derived units are newton, N; joule, J; pascal, Pa; and watt, W. The SI units for most physical quantities are:

Units

Length	meter	Electric charge	coulomb
Mass	kilogram	Electric capacitance	farad
Time	second	Electric inductance	henry
Force	newton	Magnetic flux	weber
Electric current	ampere	Temperature	Kelvin degree
Electric potential	volt	Luminous intensity	candela
Electric resistance	ohm	Luminous flux	lumen

Derived Units

Force	newton	$N = mkg/s^2$	$1.0\ N = 10^5$ dyne
			= 0.22482 lb-force
			= 3.597 ounce-force
			= 0.10197 kg-force
Work, heat energy	joule	J = Nm	$1.0\ J = 10^7$ erg
			= 0.73756 ft lb
			= 0.239 calorie
			$= 0.27777 \times 10^{-6}$ kWh
			$= 0.9478 \times 10^{-3}$ Btu
Power	watt	W = J/s	1.0 W = 0.00134 hp
			$= 10^7$ erg/s
			= 0.0569 Btu/min
			= 0.239 cal/s
			= 14.34 cal/min
			= 0.73757 ft lb/s

Derived Units (continued)

Pressure	Pascal	Pa = N/m^2	1.0 Pa = 9.9×10^{-6} atm = 0.004 in water = 0.296×10^{-3} in Hg = 0.145×10^{-3}psi = 0.0075 mm Hg

SKE Stein-Kohl-Einheiten or hard coal units or tons of coal equivalent.

SLUDGE The wet mass of waste from a sewage system. Also a general term describing the sediment from a boiler, chemical process, synfuels manufacturing, or drilling.

SLURRY FUELS Liquid and solid fuels mixed to form pumpable slurries which can be used for fuels. Examples are coal (pulverized) with oil and water and biomass materials with oil. These may be developed in the future for use with furnaces, boilers, and engines.

SMCRA Surface Mining Control and Reclamation Act.

SME Society of Manufacturing Engineers.

SNG Substitute natural gas; synthetic natural gas.

SNUPPS Standardized nuclear unit power plant systems.

SO_x Sulfur oxides, either sulfur dioxide (SO_2) or sulfur trioxide (SO_3).

SOFT COAL See Anthracite Coal; Bituminous Coal. In England soft coal is brown coal, lignite and low-Btu coals, the average calorific value of which is about one-half of most hard coals.

SOLAR Pertaining to the sun. Also see *Biomass; Energy; Photosynthesis.*

SOLAR CELL A general name for a device for converting directly the energy from the sun by solar radiation to electricity. Solar cells are of various types: Photovoltaic, thermionic, and thermoelectric. The photovoltaic cells appear to offer the greatest possibilities and are generally referred to as solar cells. Also see *Gallium Arsenide; Photovoltaic; Thermionic; Thermoelectric.*

SOLAR CONSTANT The amount of solar energy falling at normal incidence on a unit area in unit time outside the earth's atmosphere and at the mean distance from the earth to the sun. The mean value is 1.94 calories per cm^2-min or 1.94 Langley/min or 1.4 kW/m^2 (428 Btu/ft^2-hr) which varies by ± 2%. About 30% of the solar energy does not reach the surface of the earth and about 50% of the energy of the electromagnetic spectrum is in the infrared range. Also see *SERPY; Solar Radiation.*

SOLAR COOLING The use of solar energy to accomplish cooling by an absorption or gas refrigeration cycle. Solar cooling is particularly attractive because the energy from the sun is more likely available when the cooling is needed obviating the need for storage as needed for heating. Flat plate collectors do not provide a temperature high enough to obtain efficient operation of ammonia gas systems, so that considerable attention is being given to other refrigerants or solar collectors to provide higher temperatures economically. Also see *Absorption Refrigeration.*

SOLAR ENERGY The energy which is derived from the radiation of the sun which is a result of continuous thermonuclear reactions. About 70-75% of the energy intercepted by the earth goes through the atmosphere and reaches the earth. Solar energy is the primary basic energy source for the earth. Plant and animal life of the past which resulted in fossil fuel depended on solar energy. The inexhaustible, continuous source of energy from the sun is the basis for developing energy resources for the future. The term is sometimes used for that energy that comes indirectly from the sun such as wind energy and biomass energy. Also see *Biomass; Electromagnetic Spectrum; Solar Constant; Sun; Wind.*

SOLAR ENERGY COLLECTORS Various shapes and types of devices used to collect solar energy for heating. The costs and efficiencies vary greatly (Figure S-3).

SOLAR HEATING The use of energy from the sun for providing heat for space, water, and processes. The flat plate collector is normally used which provides heat below the boiling point of water. With a focusing or concentrating collector high temperatures can be obtained which can be used for cooking, processing, producing boiling water and steam which could be used for producing electric power. Passive solar heating provides an approach to maximizing the use of solar energy in the design of a building without using mechanical or electrical devices. For many solar heating applications it is desirable to incorporate methods of storing thermal energy during the day for later use at night or when the sun is not shining. Also see *Flat-Plate Collector; Focusing Collector; Heat Storage; Passive Solar Heating.*

SOLAR PANEL A name for flat-plate solar collectors used primarily for heating applications. Also see *Flat-Plate Collector.*

SOLAR POND An especially designed shallow pond with water as the principal medium for collecting and storing solar energy as heat. The features of the solar pond include: A black bottom to absorb shortwave radiation; keep water clear by using a chemical such as copper sulfate to prevent plants from growing; reducing convection of water in the pond by using a salt of sodium chloride or calcium chloride; and covering at night to reduce heat losses. Heat is

COLLECTOR					NOON HOUR	FULL DAY
SHAPE	COVER PLATE	ABSOR-BER	BACK PLATE	CHAN-NELS	%	%
	BARE	METAL	PLYWOOD	1	22	23
	FIBER GLASS	METAL	CHIP-BOARD-INSUL	1	68	63
	POLY	POLY	PLYWOOD	2	62	55
	GLASS	METAL	PLY-INSUL	2	75	68
	FIBER GLASS	DEEP-V METAL	PLY-INSUL	2	82	76
	FIBER GLASS MYLAR	DEEP-V METAL	PLY-INSUL W/REFLECT	2	89	83
	POLY	POLY	GROUND POLY	1/2	42	40
	BARE	METAL	PLYWOOD	1	14	16
	FIBER GLASS	PLY-INSUL	(W/REFLECT)	1	42	41
	FIBER GLASS	CHIP-BOARD	CHIP-BOARD	2	51	51
	POLY	13" DIA PIPE	BLACK FELT	2	48	42

Fig. S-3 Solar collector efficiencies, south facing, optimum angles, airflow of 8 cfm/ft^2.
From Kline, G. L. (1977). ASAE Paper 77-3007.

Also see *Flat-Plate Collectors; (Focusing Collectors.)*

transferred from the water in the pond through a heat exchanger to a fluid for heating applications.

SOLAR RADIATION Heat and light radiation from the sun. The total energy flux arriving at the outer boundary of the earth's atmosphere is 1.94 calories/cm^2-min = 10,320 Btu/ft^2 day. Solar radiation interacts with the air through which it passes so that the solar radiation reaching the earth's surface is reduced and the spectrum is shifted. The flux reaching a fixed location on the surface depends on latitude, season, time of day, cloudiness, air pollution, and other factors. Values averaged over a year for different locations in the United States are shown on Figure S-4. Also see *Solar Constant*.

SOLID FUEL Solid fuel for supplying heat for processing provide an alternative to electricity, gas, oil, and solar sources. Coal is the major solid fuel in the United States. Of particular interest are those fuel sources obtained from products being dried, primarily those of a cellulose base, e.g., bagasse, bark, cobs, bulbs, sawdust, shells, straw, and wood. Generally the heating value of these fuels is about 19,000 to 19,750 kJ/kg (8000 to 8500 Btu/lb) of dry material (Table S-1). The materials may be ground, chopped, shredded, or used whole. The fuel may be pelleted to facilitate handling, transportation, and storage. Combustion may be over grates or in a fluidized bed. A fluidized bed may circulate a hot inert material, such as sand, to maintain a constant flame. Initial ignition of the bed may be facilitated with a gas-fired unit. Also see *Burner; Coal, Combustion; Fuels; Gas; Heater.*

TABLE S-1
Solid Fuel Heating Values

Material	Heating value (dry product) Btu/lb	kJ/kg
Bagasse	8,000-8,700	18,600-20,235
Bark	8,750	20,350
Charcoal	12,000-13,000	27,900-30,240
Coal	9,000-14,000	20,900-32,560
Cobs, corn	8,000	18,600
Hulls, general	7,000-9,000	16,280-22,100
Hulls, cottonseed	7,100	16,515
Peat, brown	7,800-9,200	18,140-21,400
Sawdust	8,700	20,235
Straw	5,000-6,500	11,630-15,120
Wood	7,800-9,200	18,140-21,400

SOLID WASTE Waste material in solid form. About 260 million tons of municipal solid wastes are produced annually in the U.S.A., of which about 50-60% could be used for fuel. Other solid waste materials include municipal sewage waste, 12 million tons of dry matter per year in the U.S.A.; annual manure, about 250 million tons of

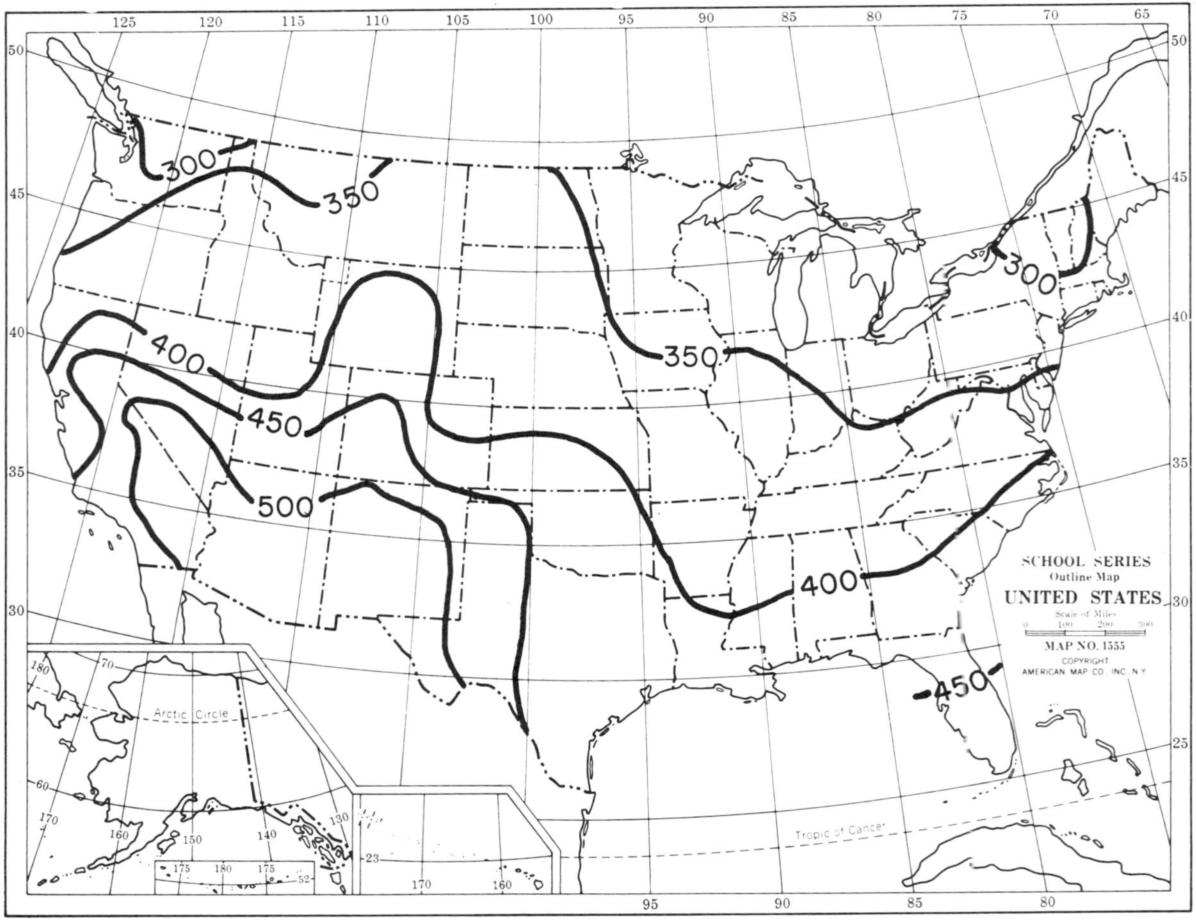

Fig. S-4 Solar radiation, mean daily values, Langleys.

organic waste in the U.S.A.; agricultural crop wastes, about 390 million tons of organic waste per year in the U.S.A.; logging and wood residues, 50 million tons per year; plus other sources. About one billion tons of organic waste are generated per year in the U.S.A. Also see *Municipal Solid Waste*.

SOLVENT REFINED COAL (SRC) Solvent refined coal is a solid fuel manufactured from coal by dissolving it in oil, partially hydrogenerating it, and filtering the product to remove particulate impurities. The process produces a nearly ash-free solid fuel. However, organic sulfur is not removed. Coal slurry is treated under pressure at 450°C with hydrogen added. Gaseous products are removed; char is produced, light oil is produced, and solvent refined coal is produced (Figure S-5). A commercial catalyst is not used. The SRC has a melting point of 150°C and low ash and low sulfur content. It is used for applications requiring a clean fuel. There are three SRC processes under development, SRC I (desulfurized solid fuel), SRC II (coal liquefaction) and SRC ½. The H-coal process is similar except that a catalyst is used. Also see *Coal Liquefaction*.

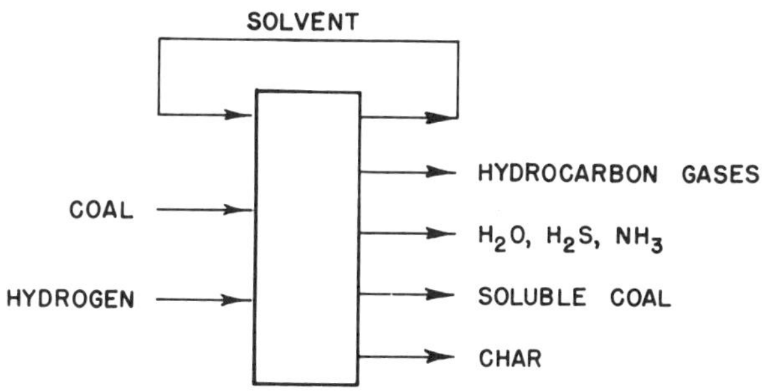

Fig. S-5 Solvent refined coal.

SORPTION See Adsorption.

SOUR GAS Natural gas which contains hydrogen sulfide, an undesirable compound for fuel or feedstock. Hydrogen sulfide can be removed with a water-alkali solution. For large quantities of hydrogen sulfide, it is removed and refined as a source of sulfur. Also see *Acid Gas; Claus Process; Natural Gas; Sweet Gas*.

SOUR OIL Crude oil which contains significant hydrogen sulfide gas. The hydrogen sulfide gas is particularly undesirable from the viewpoint of odor and as a constituent in fuel. Also see *Sweet Oil*.

SPA, SWPA Southwestern Power Administration. See Power Administrations.

SPARK IGNITION See Otto Cycle.

SPE Society of Petroleum Engineers.

SPECIFIC ENERGY Used for energy storage devices to represent the quantity of energy which can be contained or stored per unit mass, as applied to batteries and flywheels. Units used are Btu/lb, cal/g, and J/kg. Also see *Energy Density; Heat Flux Density.*

SPECIFIC GRAVITY The ratio of the density of a substance to the density of water, at a specified temperature. Also see *Apparent Specific Gravity; Density; Voids.*

SPECIFIC HEAT The ratio of the thermal capacity of a substance to the thermal capacity of water at 15°C (59°F). Another definition sometimes used is that specific heat is the amount of energy required to raise unit mass of a substance by one unit of temperature. When the units used in the latter case are calories per degree Celsius per gram or Btu per degree Fahrenheit per lb, the two definitions give the same numerical results.

SPONDUMENE A compound of aluminum, lithium, and silicate, used as the principal source of lithium. Lithium is an important element for fusion nuclear reactors.

Sr Strontium.

SPR Strategic petroleum reserve.

SPS Submerged production system.

SPURIOUS POWER Electromagnetic energy produced at frequencies different from the specified operating frequency.

SRC Solvent refined coal.

SRM Standard reference material of the National Bureau of Standards.

SRTA Spherical reflector/tracking absorber.

SSPP Solar-sea power plant.

SSPS Satellite solar power stations.

STABLE ISOTOPE A nuclide that does not spontaneously change to a different species, i.e., does not undergo radioactive decay. Also see *Radioisotope.*

STAG Steam and gas turbine combined cycle.

STANDARD ATMOSPHERE Used in meteorological work, an environment at 1 atm pressure, 760 mm (sea level) at 15°C (59°F). Also see *Conversion Factors; NTP; STP.*

STANDARDS See ANSI; ASA; ISO.

STARCH A complex carbohydrate compound of high molecular weight with the general formula of $(C_6H_{10}O_5)_x$ with the x representing the number of units in a molecule. The starch granules in vegetables determine to a large extent their texture. Starch can easily be converted to alcohol and this is a raw material from which liquid fuels can be made. Also see *Alcohols; Carbohydrates.*

STCRS Solar-thermal central receiver system.

STDS Solar-thermal distributed receiver system.

STEAM The vapor of water at or above the boiling point. Generally classified as wet steam, where all of the liquid in steam is not in a gaseous state; dry saturated steam, where all of the water is in vapor form, making 100% quality; and superheated steam, where sensible heat is added to dry steam to increase the temperature above the boiling or evaporation temperature corresponding to the pressure. The steam tables represent the pressure, temperature, and heat relationships. Steam is the most common working fluid used to convert heat energy into mechanical energy.

TABLE S-2
Saturated and Superheated Steam Properties

Pressure, psia	Temp., °F	Specific volume ft^3/lb	Heat of liquid, Btu/lb	Latent heat, Btu/lb	Total heat, Btu/lb
Saturated steam					
14.7	212.0	26.8	180.1	970.3	1150.4
20	227.9	20.1	196.2	960.1	1156.3
30	250.3	13.7	218.8	945.3	1164.1
40	267.3	10.5	236.0	933.7	1169.7
50	281.0	8.5	250.1	924.0	1174.1
60	292.7	7.2	262.1	915.5	1177.6
70	302.9	6.2	272.6	907.9	1180.5
80	312.0	5.5	282.0	901.1	1183.1
90	320.3	4.9	290.6	894.7	1185.3
100	327.8	4.4	298.4	888.8	1187.2
125	344.3	3.6	315.7	875.4	1191.1
150	358.4	3.0	330.5	863.6	1194.1
200	381.8	2.3	355.4	843.0	1198.4
250	400.9	1.8	376.0	825.1	1201.1
300	417.3	1.5	393.8	809.0	1202.8
400	444.6	1.4	424.0	780.5	1204.5

STEAM AND GAS TURBINES (STAG) A combined-cycle power system using steam and gas turbines, usually to produce electricity (Figure S-6). STAG plants have as a potential using coal-derived fuel (CDF) and other synthetic fuels. Also see *Integrated Gasification Combined Cycle (IGCC)*.

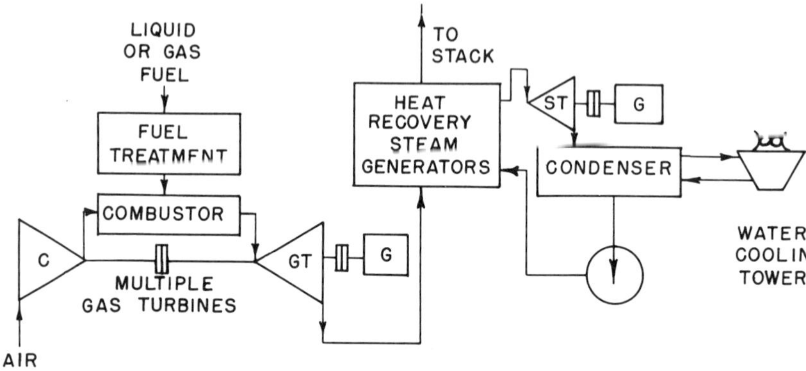

Fig. S-6 Steam and turbine combined cycle *(Corman, 1979)*.

STENOTHERMIC Those reactions which are tolerant of a very narrow range of temperatures.

STILLAGE The mash or residue remaining after grain is fermented to produce alcohol. The stillage can be fed to livestock. Also called distillers grain. Also see *Distillers Grain*.

STIRLING ENGINE An external combustion engine which operates on the Stirling cycle. There are two basic forms of Stirling engine. One employs two pistons moving in a single cylinder and connected to a shaft in such a way that their motions are 90° out of phase. The other uses two cylinders, each containing a piston connected to a common shaft and again with piston motions 90° out of phase. Each type of engine draws heat from a hot source and delivers heat to a cold sink. The working fluid in a Stirling engine is contained in a closed volume consisting of two connected spaces. As the engine operates gas moves from one space to another. An important part of the device is a regenerator, located in the channel connecting the two partial volumes. The regenerator stores and releases heat as the gas moves back and forth. The Stirling engine has good efficiency and offers a means of utilizing low-temperature heat (Figure S-7).

STOICHIOMETRY The calculation or the attainment of the exact proportions of constituents required for a chemical reaction. The principles of stoichiometry may be used to study an energy system. Stoichiometric combustion is complete combustion in which the amount of oxygen available is exactly sufficient to oxidize the fuel. Also see *Combustion*.

$$CH_4 + 2\ O_2 \rightarrow CO_2\uparrow + 2\ H_2O + \text{heat.}$$

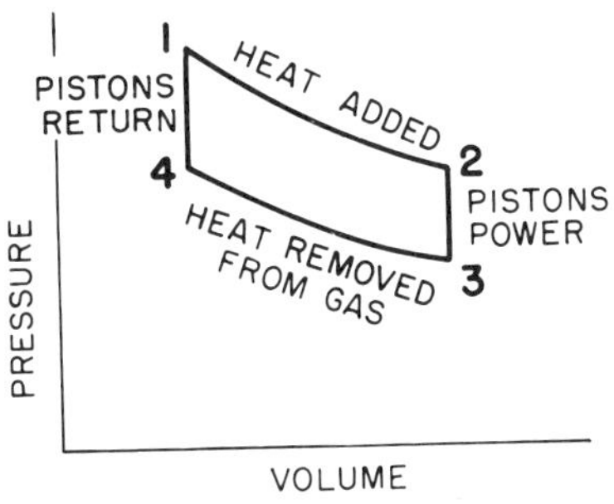

Fig. S-7 Stirling cycle.

STOPE An underground excavation from which minerals or ores are removed in horizontal steps.

STORAGE BATTERY See Battery.

STP Standard temperature and pressure. Also see *NTP*.

STRAIGHT CHAIN HYDROCARBONS Those hydrocarbons which have a configuration in which the carbon atoms are connected in a chain without branching. Diesel fuel has a large portion of straight hydrocarbons in contrast to gasoline fuel which has a high proportion of branched chain hydrocarbons. The aliphatic hydrocarbons have a straight chain, and indeed the members of the alkane series, C_nH_{2n+2}, in the normal form are straight chains. Isomers of the hydrocarbons are not straight chain. Also see *Aliphatic Hydrocarbons; Alkane; Saturated Hydrocarbons*.

STRATEGIC PETROLEUM RESERVE (SPR) A large volume of crude oil stored in underground chambers near the Gulf of Mexico. Because of the dependence of the U.S.A. on foreign crude oil, crude oil is being purchased by the government and placed in storage as a precaution against possible interruptions in supplies. The goal is to have one billion barrels of oil in the reserve. The rate of filling the SPR was 100,000 bbl per day in 1980, although the plan has been to fill at the rate of 300,000 bbl per day. Also see *Energy Security Act*.

STRATOSPHERE See Atmosphere.

STRIPPED GAS See Natural Gas.

SUBBITUMINOUS COAL A rank of coal between bituminous and lignite. It includes brown coal in some classifications and has a calorific value ranging from 19,250 to 30,150 kJ/kg (8300 to 13,000 Btu/lb). Vast deposits of low sulfur bituminous coal are found in Montana and Wyoming. They are used primarily as fuel in coal-fired power plants. In some classifications, also called black lignite. Also see *Bituminous Coal; Lignite*.

SUBSTITUTE NATURAL GAS A name often used, perhaps misleadingly, for synthetic natural gas (SNG).

SUGAR Any of a number of sweet carbohydrates. The chemical formulae for sugars are of the form $C_nH_{2n}O_n$ or, $C_n(H_{2n-2}O_{n-1})$. Sugars are converted to alcohols by the action of yeast and thus are potential sources of fuel. Also see *Carbohydrates*.

SULFUR A non-metallic chemical element, which has atomic number 16 and atomic weight 32.1. It is a constituent of many minerals and fossil fuels. The smelting of sulfur-bearing ores, the burning of sulfur-containing fuels, and the manufacture of various products generate sulfur dioxide, an air pollutant that produces a number of damaging effects including offensive odors, lung irritation, and acid rain. In the atmosphere sulfur dioxide undergoes further chemical transformation partly to sulfuric acid and partly to particulate sulfates. It is these secondary pollutants that are directly responsible for many of the environmental effects of sulfur dioxide. Manufacturing industries, smelters, oil refineries, and electric utilities have installed equipment to remove the sulfur from the raw material, the intermediate products, or the exhaust gases of the processes involved so as to reduce the environmental impacts of sulfur. Also see *Claus Process; Desulfurization; Sour Gas; Sour Oil*.

SULFUR DIOXIDE One of the undesirable products of combustion, in approximately the following magnitude (*Junge, 1980*):

Fuel	lb/MM Btu	Ng/J
Wood	0.17	73
Liquid	0.8	340
Solid fossil	1.2	520

Also see *Sulfur*.

SUN The central body in the solar system. Data describing the sun:

Average radius	$(6.96 \pm .01) \times 10^5$ km
Mass	$(1.991 \pm .002) \times 10^{30}$ kgm
Average density	$1.410 \pm .002$ g/m^3
Equatorial rotation period	$(2.125 \pm ?) \times 10^6$ sec (solar seconds)
Gravitational acceleration	$(2.7372 \pm 7.0) \times 40^4$ cm/sec^2
Solar constant (solar surface)	8.87×10^4 cal/cm^2/min
Solar constant (earth mean distance)	1.920 cal/cm^2/min
Sun to earth distance (greatest)	$(1.5207 \pm .0007) \times 10^8$ km
Sun to earth distance (smallest)	$(1.4707 \pm .0007) \times 10^8$ km

The sun is a self-luminous hot ball of fire consisting primarily of helium and hydrogen undergoing a thermonuclear reaction. The sun is a star. The center of the sun has a specific gravity of 96 and is at a temperature of 20 million degrees Celsius while the surface is less dense than atmospheric air and is at 6000°C (10,832°F). The sun produces 3.9×10^{26} J/s of energy of which the earth intercepts 1.75×10^{17} J or 1.52×10^{18} kWh per year. Seventy to 75% of the intercepted energy penetrates the atmosphere and is received on the earth's surface. A stable life of 10^{10} yr is estimated for the sun. Also see *Solar Energy.*

SUNDANCE WELLS Hot water wells which produce hot water at a temperature which can be used for space heating.

SUPERHEAT The heat added to saturated vapor to increase the temperature of the vapor above the vaporization temperature corresponding to the pressure of the vapor. Also see *Superheated Steam.*

SUPERHEATED STEAM Steam to which heat has been added to increase the temperature above that temperature corresponding to the temperature of boiling for the pressure of the vapor. By adding sensible heat to dry saturated steam, while maintaining the same pressure, superheated steam is obtained (Table S-3). Each degree increase in temperature is identified as a degree superheat. Heaters for vapor heat transfer, known as superheaters, are used. The greater the degree of superheat, the greater the amount of heat that can be removed without condensation. Also see *Steam.*

SWEET GAS Natural gas which contains small amounts of hydrogen sulfide and carbon dioxide gases as contrasted to sour gas. Also see *Acid Gas; Sour Gas; Sour Oil.*

SWEET OIL Crude oil which contains small amounts of hydrogen sulfide and carbon dioxide gases as contrasted to sour oil. Also see *Sour Oil.*

SYNCRUDE See Synthetic Crude Oil.

SYNFUEL A synthetic fuel, a manufactured fuel, which has the same properties as a common natural fuel. Examples are synthetic natural gas and synthetic gasoline.

SYNGAS See Synthetic Gas.

SYNTHANE A coal gasification process applicable to a wide variety of coals. Coal is exposed to an oxygen-steam environment in a pressurized chamber in a fluidized bed. A synthesis gas ($CO + H_2$) is produced which can be upgraded to pipeline quality. Also see *Synthesis Gas.*

SYNTHESIS GAS A gas consisting primarily of a mixture of carbon monoxide and hydrogen and produced by reaction of hydrocarbons or carbohydrates with oxygen and steam. Following the production of synthesis gas other processes are usually carried out to remove

TABLE S-3
Superheated Steam

Pressure, psia		Temperature, °F 300	400	500
	sh	88.00	188.00	288.00
14.7	v	30.52	34.65	38.75
(212°F)	th	1192.0	1238.9	1285.4
	sh	49.66	149.66	249.66
30	v	14.82	16.89	18.92
(250°F)	th	1189.2	1237.4	1284.2
	sh	25.55	125.55	225.55
45	v	9.79	11.20	12.57
(274.5°F)	th	1185.9	1235.8	128.30
	sh	7.29	107.29	207.29
60	v	7.26	8.35	9.39
(292.7°F)	th	1181.8	1234.0	1281.8
	sh		92.4	192.40
75	v		6.64	7.49
(307.6°F)	th		1232.2	1280.6
	sh		79.7	179.73
90	v		5.5	6.22
(320.3°F)	th		1230.0	1279.3
	sh		68.6	168.62
105	v		4.69	5.32
(331.4°F)	th		1227.6	1278.0

sh = superheat, °F; v = specific volume, cu ft/lb; th = total heat, Btu/lb

Conversion factors: *1 psi = 6.8947 kPa = 6.8947 x* 10^3 *N/m*^{2}
1 ft^3*/lb = 6.243 x* 10^{-2} *m*^{3}*/kg*
1 Btu/lb = 2.326 x 10^3 *J/kg*

Based on Van Wylen, G. J., and Sonntag, R. E. (1976). Fundamentals of Classical Thermodynamics. John Wiley and Sons, New York, NY.

impurities or to upgrade the gas to a higher heat content. Also see *Methanation*.

SYNTHETIC CRUDE OIL A synthetic liquid similar to crude oil but produced from coal; also known as syncrude. Synthetic crude can be produced directly from coal or from the product of a gasification process. The major approaches to coal liquefaction include: Pyrolysis, high pressure hydrogenation; solvent refining, and upgrading of gas made from coal to crude oil equivalent. The COED process produces about 1.55 bbl of crude per ton of soft coal, plus 8000 to 10,000 SCF of off-gas. Also see *Bergius; Coal Liquefaction; COED; Consol; H-Coal; SRC; Synthoil*.

SYNTHETIC FUELS CORPORATION A U.S.A. government supported organization created under the Energy Security Act (1980) to develop a new industry for producing synthetic fuels. The goal is to produce 500,000 bbl/day of oil or equivalent by 1987 and 2,000,000 bbl/day of oil or equivalent by 1992. Also see *Energy Security Act; Strategic Petroleum Reserve.*

SYNTHETIC NATURAL GAS; SUBSTITUTE NATURAL GAS (SNG) A synthetic gaseous fuel containing a high proportion of methane. SNG made from coal is expected to be an important future fuel in the U.S.A.

SYNTHOIL A process developed by USBM for liquefying coal to produce fuel oil using a high temperature of 450°C (842°F) and a high pressure of 150-300 atm. The process involves the addition of hydrogen in the presence of a catalyst. A yield of 90% of fuel oil from the coal is claimed. Also see *Coal Liquefaction.*

SYNTHOL A process developed by SASOL for converting coal to gas using heat from the Fischer-Tropsch process. The process is carried out in a reactor at 160°C (320°F) at a pressure of 20 atm using a catalyst at a higher temperature.

T

TANK FARM   An area on which several storage tanks for crude oil or petroleum products are placed.

TANKERS   Ships constructed to transport liquids across oceans or large lakes. Tankers for handling the international oil and liquefied natural gas trade have become progressively larger. There is an economy of scale which can be achieved by going to larger tankers. The cost of operation of ULCC's per unit volume is only about one-fifth that for the GP vessel. The classification of crude oil tankers is as follows:

General purpose, GP	17,000-24,999 tons
Medium range, MR	25,000-44,999
Larger range, LR 1	45,000-79,999
LR 2	80,000-159,000
LR 3	160,000-319,999
Ultra large crude carrier, ULCC	above 320,000 tons by one scheme
Very large crude carriers, VLCC	above 160,000 tons

TAO   Technology Assistance Offices (SBA).

TAPPI   Technical Association of the Pulp and Paper Industry.

TAR   A heavy, dark viscous liquid product of distillation, refining or carbonization of hydrocarbons, coal, peat, shale, wood, and other organic products. The term tars may be applied to heavy crude oils up to an approximate API gravity of 10. Also see *Asphalt; Char; Gilsonite; Pitch.*

TAR, BITUMINOUS   A dark colored viscous fluid insoluble in water but soluble in carbon disulfide. Pitch is produced by distillation of bituminous tar. Pitch is taken off the bottom of the refinery of crude petroleum.

TAR, COAL The hydrocarbon distillates produced by the destructive distillation of coal. The hydrocarbons are largely unsaturated.

TAR DISTILLATES The distillates which are the product of distillation of tar or the bottom fractions from cracking crude petroleum.

TAR OILS The oils which are produced during the distillation of tar. The tar oils could include naphtha, light oil, and creosote. Also see *Creosote; Light Oil; Naphtha.*

TAR SANDS Sub-surface deposits of sedimentary formations of sand impregnated with heavy petroleum liquid or solid hydrocarbons (bitumens). The DOE definition of tar sands is a native asphalt, solid and semi-solid bitumen that will not flow to a borehole under natural reservoir conditions of temperature and pressure or having a viscosity exceeding 15,000 centipoise under standard conditions at the surface. Vast deposits of tar sand exist in Athabasca, Alberta, Canada. The raw material is removed by strip mining. Oil is obtained by heating to 80°C (175°F) to separate which is followed by additional processing. One-half barrel of oil is obtained per tonne. U.S.A. tar sands contain up to 18% bitumen by weight. Also see *Athabasca; Bitumen; Oil Sand.*

TASMITE See Bituminous Shale.

TBOE Thousand BOE.

TCE Tons coal equivalent.

TCF, Tcf, tcf Trillion cubic feet or teracubic feet, 10^{12} cu ft.

Tcf/y Trillion cubic feet per year.

T/D, TPD Tons per day.

TDS Total dissolved solids.

TEL Tetraethyl lead. See Lead; Octane Number.

TEMPERATURE The property that quantifies the common notions of hotness and coldness of an object. A more fundamental but abstract thermodynamic definition of temperature can be given as the rate of change of internal energy of a system with respect to increases in its entropy assuming that other independent extensive variables remain constant. The thermodynamic definition can be made operational using the relation between the heat absorbed (Q_1) and heat rejected (Q_2) by a reversible heat engine (Carnot cycle) operating between reservoirs at temperatures T_1 and T_2. The thermodynamic temperature scale can be constructed using $Q_1/Q_2 = T_1/T_2$.

Various thermometers have been devised to measure temperature. One of the most accurate, the gas thermometer, uses the pressure of a gas to indicate temperature.

The four common temperature scales are:

	Absolute zero	Freezing point of water	Boiling point of water
Kelvin, K	0	273.2	373.2
Celsius, °C	-273.2	0	100
Rankine, R	0	491.6	671.6
Fahrenheit, °F	-459.6	32.	212.

TENNESSEE VALLEY AUTHORITY An independent corporate agency of the U.S.A. government established in 1933 to develop resources of the Tennessee River and related areas. One of the major resources developed is for producing electricity. Originally development was based on hydroelectric facilities with 42 dams in the system (with a capacity of 4 million kW) but has more recently been building coal-fired steam plants (15 million kW) and nuclear power plants (8.5 million kW).

tEP Tons equivalent petrol (England).

TERA Total energy resource analysis.

Tera (T) A prefix meaning 10^{12}. Also see *Trillion*.

TERTIARY OIL RECOVERY The practice of extracting additional oil or gas from a reservoir beyond secondary recovery. Thermal or chemical methods are used for tertiary removal such as steam injection, *in situ* combustion, and solvents. Some authors include steam injection under secondary oil recovery. Up to 25% of the total content might be removed by one or more of these methods. Also see *Enhanced Oil Recovery; Oil Recovery; Secondary Oil Recovery*.

TES Thermal energy storage.

TFTR Tokamak Fusion Test Reactor (Princeton).

Th Thorium.

THERM In English units, may refer to the heat equivalent of 100,000 Btu. In older literature, therm was another name for the British thermal unit or calorie. One therm = 25,200 kcal = 25.2 Mcal. Also see *Q; Quadrillion; Quintillion; Thermie*.

THERMAL A word pertaining to heat or temperature. Also see *Thermo-*.

THERMAL CAPACITY The heat capacity of a body on the basis of amount of heat required to raise the temperature 1°C. Also see *Specific Heat; Thermal Conductivity*.

THERMAL CONDUCTION A process in which heat is propagated as a result of the thermal motion of the elementary particles of the material (molecules, atoms, free electrons). Essentially, conduction occurs in solid bodies and in very thin, stationary layers of liquid or gas. Also see *Thermal Conductivity*.

THERMAL CONDUCTIVITY A measure of the movement of heat through substances by kinetic molecular activity. Thermal conductivity, k, has units of cal/cm-sec-°C in the cgs system and Btu-in/hr ft^2 °F or Btu/hr ft °F in the English system and J/ms °K or W/mK in SI units. A conversion factor is 1 cal/cm sec °F = 0.80620 Btu-in/ft^2sec °F = 418.7 J/ms °K = 418.7 W/mK. The rate of heat transfer, q, Btu/unit time, by conduction,

$$q = \frac{k}{x} (t_2 - t_1)A$$

where

k = thermal conductivity
x = distance perpendicular to the face
A = the cross sectional area
$t_2 - t_1$ = the temperature difference

These relationships are used for steady-state heat transfer. Also see *Conduction*.

THERMAL CONVECTION The transfer of heat due to movement of a fluid. Movement occurs because of differences in density or temperature in natural convection and because of forced agitation in forced convection. The overall heat transfer across a boundary between a solid and a fluid is strongly influenced by the heat transfer properties of a film or layer of fluid adjacent to the heat transfer surface. Heat moves through the film by conduction and the film transfer rate in turn controls the rate of heat transfer by convection. The film coefficient or surface thermal conductance h (represented with f by some authors) in W/m^2K or Btu/hr ft^2 °F is used to calculate the heat transfer by convection.

$$q = h\,A\,(t_2 - t_1)$$

Also see *Convection Heat Transfer*.

THERMAL DIFFUSIVITY The rate at which a substance with a nonuniform temperature approaches equilibrium based on thermal conductivity, k, specific heat, c, and the density, rho, ρ:

$$\alpha = k/c\rho$$

The thermal diffusivity, α, is used in calculations of unsteady state heat conduction problems. Calculations of unsteady-state heat transfer (where the difference in temperature varies with time), are usually made using prepared graphs or charts with thermal diffusivity as one of the parameters. Also see *Diffusion; Thermal Conductivity*.

THERMAL CRACKING A breaking of carbon-carbon bonds in hydrocarbons and carbohydrates accompanied by a reduction in molecular size, with the formation of aromatics. Cracking is accomplished by the action of heat at temperatures of 450° to 510°C (842° to 950°F) in a vacuum.

THERMAL GRADIENT As related to the crust of the earth, the increase in temperature with depth, which is approximately 25°C/km.

THERMAL RADIATION The energy emitted by all bodies (at a temperature above absolute zero) as a result of transitions among excited molecular states. Energy is transferred from one body to another by thermal radiation without heating the space between, except when the space has substances which absorb energy. Thus, heat transfer by radiation is determined by the temperatures of the emitting and receiving bodies. The Stefan-Boltzmann constant, σ, is used in thermal radiation calculations. The heat loss from a black body with emissivity $\varepsilon = 1$ is

$$q = \sigma A T^4.$$

THERMAL REFORMING The upgrading of a product of refining, such as distillates for gasoline, by using heat at a temperature of 550°C (1020°F) or above and at a high pressure such as 100 atmospheres. For gasoline, catalytic reforming has replaced thermal refining. Also see *Catalytic Reforming; Reforming*.

THERMAL SPLIT In geothermal heating, the separation of hot rock as a result of an explosive charge to permit a fluid such as water or liquid sodium to pass through the openings and be heated.

THERMAL STORAGE See Heat Storage.

THERMIC A word which refers to heat.

THERMIE The amount of energy required to raise the temperature of one tonne (metric) of water by 1°C. 1 thermie = 1 Mcal = 1000 kcal = 0.03968 thermie. Also see *Therm*.

THERMIONIC CELL A device which directly converts high temperature heat energy into electricity. High temperatures of at least 1200°C (2190°F), such as those which could be obtained on a focusing solar collector, are required. The thermionic cell is similar to a vacuum tube in which a heated cathode emits electrons to an anode to produce an electric current. The cost of electricity produced by this method appears to limit its immediate use for commercial purposes. Also see *Photovoltaic; Solar Cell*.

Thermo- A prefix used for thermal or heat. Also see *Thermal*.

THERMOCOUPLE The sensing part of an instrument for measuring temperature in which a pair of two dissimilar metal wires are joined. Junctions are made with iron-constantan, copper-constantan, alumel-chromel, etc.

THERMODYNAMICS The subject that describes the processes and equilibrium states of a system in terms of a set of observable quantities such as temperature, pressure, volume, internal energy, entropy, and heat flow. Thermodynamics is widely used in chemistry, physics, biology, and engineering to help describe chemical and biological reactions, physical processes, and performance of heat engines. Also see *First Law of Thermodynamics; Second Law of Thermodynamics; Third Law of Thermodynamics; Zeroth Law of Thermodynamics*.

THERMOELECTRIC Referring to a method of converting heat to electricity with a thermocouple in which two dissimilar metals are joined, with a hot and a cold junction at different temperatures providing for a flow of electrons. With hot and cold junctions in the ambient range 1% or less efficiency is obtained, but with a hot junction temperature of 750°C (1380°F), 6% efficiency of conversion may be obtained.

THERMONUCLEAR REACTION A nuclear reaction that occurs as a result of the thermal interaction of reacting nuclei. The particle reactants must be in thermal equilibrium before the reaction is considered to be thermonuclear.

THERMOPHILIC A term meaning "heat loving" applied to bacteria which grow at higher than normal temperatures in the range 40 to 71°C (100 to 160°F) at an appropriate moisture content. Thermophilic organisms can grow in hot springs and/or heat exchangers.

THERMOPILE A group of several thermocouples connected to a galvanometer which is calibrated to give the temperature. Thermopiles are widely used for solar and other radiation measurements. Also see *Thermocouple*.

THERMOSPHERE The portion of the atmosphere from 80 km outside the earth extending to outer space. Also see *Atmosphere*.

THIRD LAW OF THERMODYNAMICS There are several statements of the law:

1. It is not possible to attain the absolute zero of temperature by a finite series of processes.

2. The entropy change associated with any isothermal reversible process of a condensed system approaches zero as the absolute temperature approaches zero.

3. The entropy of a system cannot be reduced to its zero point value by a finite series of processes.

Another frequent statement of the Third Law, closely related to (2) above, is that the entropy of all pure condensed systems can be taken to be zero at the absolute zero of temperature.

THORIUM (Th) A naturally occurring radioactive heavy metal with atomic number 90 and atomic mass 232. Thorium is an important nuclear energy material. It is probably more common than uranium on the earth's crust. When it absorbs a neutron, it undergoes a series of nuclear reactions that convert it to the nuclear fuel ^{233}U. There is a reactor concept in the U.S.A., the HTGR, that could use the ^{232}Th - ^{233}U fuel cycle once an inventory of ^{233}U is built up. ^{232}Th decays by alpha emission with a half-life of 14 billion years. Neutron absorption in ^{232}Th converts it to ^{233}Th which decays first to ^{233}Pa and then to ^{233}U. The uranium isotope, ^{233}U, is a nuclear fuel.

THOUSAND FEET BOARD MEASURE (Mbm) One thousand board feet. One board foot is the volume of a space one foot long, one foot wide, and one inch thick. One board foot has a volume of 144 in^3. One thousand feet board measure, or 1000 board-feet, has a volume of 144,000 in^3 or 2.36 m^3. Also see *Cord*.

THREE MILE ISLAND (TMI) An island in the Susquehanna River near Harrisburg, PA., on which are located two nuclear reactors operated by Metropolitan Edison Company. One of these two reactors, Three Mile Island, Unit 2, experienced a serious accident in March, 1979. The term TMI is often used to refer to the accident itself.

THROUGHPUT The volume of feedstock charged to processing equipment, either per unit time for a continuous operation, or per batch for a batch operation.

THTR Thorium high-temperature reactor.

TJ/y Terajoules per year.

TIDAL ENERGY Kinetic and potential energy associated with the tides. Tidal energy can be converted to mechanical and subsequently electrical form in several ways. The most common is to utilize a single natural basin that experiences tidal flows. The mouth of the basin is blocked by a dam that contains sluice gates and hydroelectric turbogenerators. In operation water enters the basin through open sluice gates. At high tide the gates are closed. When the tide ebbs sufficiently water is released through the hydroturbines to generate electricity. In a more elaborate system, currently in operation at the mouth of the Rance River in France, horizontal axis reversible turbines are used so that electricity can be generated during both incoming and outgoing tides. If two adjacent basins are used rather than one, power can be generated continuously. One basin is kept filled by high tides, the other empty by low tides. The two are connected by channels containing hydropower units which can be operated continually as a result of the level difference between the two basins. Tidal energy is most practical where local topography provides large tidal ranges (high to low variation in water level) and suitable natural basins. Examples are the Bay of Fundy on the Atlantic Coast at the U.S.A.-Canadian border where the average tidal range over the year is as much as 10.7 m in some places and along the French coast where the average range is 8.4 m at some places. The highest tide is 16 m (52.5 ft) and occurs in the Bay of Fundy. The peak electricity is produced from tides each 12 hr, usually not in concert with the demand. Also see *Pump Storage*.

TIME See Cosmic Year; Geologic Time.

TIME CONSTANT (lag coefficient, relaxation coefficient) A constant (τ) describing the time required for some physical quantity to change from an initial value Q_0 to a final value Q_∞ (Figure T-1). τ is the time required for $Q-Q_\infty$ to reach e^{-1} of its initial value Q_0-Q_∞. The value of Q at any time is given by

$$Q = Q_\infty + (Q_0 - Q_\infty)e^{-t/\tau}$$

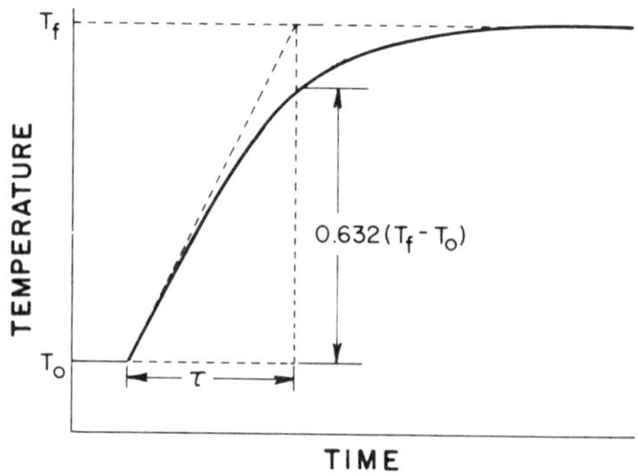

Fig. T-1 Time constant.

Tl Thallium.

TMI Three Mile Island.

TML Tetramethyl lead. See Lead; Octane Number.

TMS The Metallurgical Society of AIME.

TNC Transnational corporation.

TOE Metric tons of oil equivalent (in energy content).

TON; TONNE A measure of weight, which in the U.S.A. is 20 hundredweight (20 cwt), where 1 cwt = 100 lb. One ton is 2000 lb, also known as the short ton. In England the ton of 20 British hundredweight, where 1 cwt = 112 lb, is 2240 lb, also known as the long ton or gross ton. The metric ton, which is 1000 kg or 2204.6 lb, is represented by tonne in the SI terminology.

TOPPING The procedure of removing from crude oil the light components such as gases and gasoline, by distillation. The remaining product is called topped crude oil.

TOPPING CYCLE A system that converts heat to mechanical energy and that is installed ahead of the main unit in order to increase the fraction of heat converted to mechanical or electrical energy. When used in reference to a conventional boiler-steam turbine combination, a topping cycle might consist of a gas turbine or a magnetohydrodynamic converter. Also see *Bottoming Cycle.*

TOSCO The Oil Shale Corporation.

TOSCOAL A process of The Oil Shale Corporation (TOSCO) to convert low-Btu coal to higher grade materials which can be used in conventional combustion systems. Pyrolysis, using heated ceramic balls in a rotating drum at 538°C (1000°F), is the major operation. Char, tar, and gas are the major products of pyrolysis.

TOTAL HEAT The sum of the sensible heat and the latent heat in a substance; in air-vapor mixtures the total heat is also called enthalpy. Also see *Enthalpy.*

TOWN GAS A general name formerly used to describe a medium-Btu gas made from coal and supplied to cities and towns. Also called coal gas. The pipelines were originally used to deliver town gas and have been greatly expanded to provide natural gas to consumers. Also see *Coal Gas.*

TPD Tons per day.

TRANSMISSIVITY; TRANSMITTANCE; TRANSMISSION COEFFICIENT; TRANSMISSIBILITY The fraction (τ) of radiation energy falling on a surface which is transmitted through the body. Other related quantities are the fraction reflected, ρ, the reflectivity, and the fraction absorbed, α, the absorptivity. Note that $\alpha + \rho + \tau = 1$. Most gases have a high value of τ.

TRANSURANIC WASTES (TRU) Wastes containing more than 10 nanocuries of transuranic activity per gram. Transuranic wastes are byproducts of nuclear fuel reprocessing, plutonium fuel processing, and nuclear weapons manufacture. About 370,000 m^3 of these materials are stored in the U.S.A. *(Gloyna, 1980).* Also see *High-Level Wastes; Low-Level Wastes.*

TRI-GENERATION A procedure for utilizing an energy source, such as coal, which is gasified in the first step and is integrated with co-generation to obtain optimal thermal energy utilization. Also see *Co-Generation*.

TRILLION 10^{12} often represented by T, which also means tera. Also see *Tera*.

TRILLION CUBIC FEET (Tcf) A measure of volume used for large quantities of gaseous fuel. The U.S.A. uses approximately 20 Tcf of gas annually (1980).

TRITIUM A radioactive isotope of hydrogen having atomic number 1 and atomic mass number 3. Tritium emits low energy beta rays and decays with a half-life of 12.3 years. It is a potential fuel in the fusion process but also is a potential radiation hazard in the operation of both fission and fusion reactors. Also see *Deuterium; Heavy Water; Lithium*.

TROPOSPHERE See Atmosphere.

TRU Transuranic radioactive wastes.

T-s DIAGRAM A graphical representation, usually with T (absolute temperature) on the ordinate and s (entropy) on the abscissa. The area under the curve of a reversible process represents the heat transfer during the process. Also see *i-x Diagram*.

TSCA Toxic Substances Control Act.

TSP Total suspended particulates.

TURBINE A device for changing the thermal or hydraulic energy of a flowing fluid to mechanical power of a rotary shaft. The turbine is used to drive machines of many different types, the most important of which is for driving an electric generator. The types are steam turbine, gas turbine, hydraulic turbine, and wind turbine (windmill). Also see *Bulb Turbine; Gas Turbine; Windmill*.

TVA See Tennessee Valley Authority.

TWh Terawatt-hours, 1 TWh = 10^9 kWh = 10^{12} Wh.

U

^{233}U, ^{235}U, ^{238}U The three isotopes of uranium that are most important in nuclear fission power. ^{233}U and ^{235}U are nuclear fuels; i.e., have large fission cross sections. ^{238}U is a fertile material which is transformed into ^{239}Pu through a two-step radioactive decay following neutron absorption. Also see *Fertile Material; Uranium.*

UCS Union of Concerned Scientists.

UHV Ultra-high Voltage.

UKAEA United Kingdom Atomic Energy Authority.

UL Underwriters' Laboratory, Inc., 207 E. Ohio Street, Chicago, IL 60611.

ULCC Ultra Large Crude Carrier. See Tankers.

ULTRA LARGE CRUDE CARRIER (ULCC) An oil tanker carrying over 350,000 long tons (2240 lbs) when loaded. Also see *Tankers.*

ULTRAVIOLET RAY Short wavelength electromagnetic radiation just beyond the visible spectral range. The ultraviolet wavelength range lies between 4×10^{-5} cm and 1.8×10^{-5} cm. Ultraviolet rays are needed by the body to make Vitamin D, and will produce fluorescence of some materials. Also see *Electromagnetic Spectrum.*

ULTIMATE RECOVERABLE RESERVES The total quantity of fuel or mineral estimated to be recoverable from an ore body or reservoir using economical procedures. Also see *Resources.*

ULTRA-HIGH VOLTAGE (UHV) See Voltage Classification.

UNAVAILABLE ENERGY Heat that cannot be converted into work during the operation of a heat engine. A cyclic engine that withdraws an amount of heat, Q, from a high temperature reservoir at absolute temperature, T_H, uses it to do work, and rejects waste heat to a low temperature reservoir at temperature T_L cannot convert all of Q into work. The Second Law of Thermodynamics requires that at least

$$\frac{T_L}{T_H} Q$$

be rejected to the low temperature reservoir no matter how perfect the engine so long as its operation is cyclic and the amounts of heat withdrawn and rejected are evaluated for an integral number of cycles. The energy

$$\frac{T_L}{T_H} Q$$

is unavailable for conversion to work. Also see *Entropy; Exergy.*

UNDERGROUND TRANSMISSION SYSTEM An electric transmission line that is located underground. Underground electric transmission systems are installed in place of overhead lines where they are unsafe or unsightly. Underground systems include the following types: High-pressure oil-filled, force cooled, compressed gas insulation, resistive cryogenic, and superconducting cables.

UNION OF CONCERNED SCIENTISTS (UCS) An organization of scientists dedicated to informing the public of the hazards of nuclear power. Located at 1208 Massachusetts Avenue, Cambridge, MA 02138.

UNIT OPERATIONS The study of physical changes which occur during processing of materials, and the design of equipment and systems based on these studies. Typical subjects for unit operations studies are drying, dehydration, heat transfer, and distillation.

UNIT PROCESS The study of chemical changes which take place in materials subjected to processing and handling in reactions such as combustion, oxidation, and polymerization. Unit process studies are used to determine the best reaction conditions as a basis for design of processes and associated equipment.

UNIVERSAL CONSTANTS See Constants.

UNSATURATED HYDROCARBON A compound of carbon and hydrogen with a molecular structure that includes double or triple carbon bonds. Also see *Saturated Hydrocarbons.*

UNSTABLE ISOTOPE A nuclide that undergoes radioactive decay. Also see *Radionuclide.*

U_3O_8 Uranium oxide, the international standard basis on which uranium concentrate is marketed.

URANIUM A naturally occurring radioactive element with atomic number 92. The naturally occurring isotopes are ^{235}U, with half-life 0.71 billion years, and ^{238}U, with half-life 4.5 billion years. Natural uranium is 99.3% ^{238}U and 0.7% ^{235}U. The isotope ^{235}U is the nuclear fuel used in light water reactors. The isotope ^{238}U is a fertile material which, following neutron absorption, converts to ^{239}Np and then to the nuclear fuel ^{239}Pu. The uranium isotope ^{233}U also is a nuclear fuel. It does not occur in nature but can be produced by neutron absorption in ^{232}Th. Also see *U*.

URANIUM PROCESSING The processing which converts uranium ore into nuclear reactor fuel. Also see *Nuclear Fuel Cycle*.

USASI U.S.A. Standards Institute.

USBM United States Bureau of Mines.

USDA United States Department of Agriculture.

USDC United States Department of Commerce.

USDI United States Department of Interior.

USGS United States Geological Survey.

USLE Universal soil loss equation.

UTILITY COST OPTIONS Relative costs of electricity produced from coal and synfuels by various systems (1980):

	Relative cost
Coal purchased	
Conventional plant/FGD	184
Integrated gasification combined cycle (IGCC)	149
Synfuels purchased	
High-Btu gas/cc	179
Medium-Btu gas/cc	126
Syncrude/cc	170
Methanol/cc	188

Todd, D. M., and Zeltmann, C. E. (1980). Implementation of the fuel use act. *Electric Forum 6(2):3-7.*

U-VALUE The overall heat transfer coefficient or thermal conductance of a building wall, floor, ceiling, roof, or other element; of piping or tubing carrying heating or cooling fluids; or of heat transfer equipment. The U-value is normally applied to a cross-section of more than one material and considers the film conductance and air. The total heat transfer can be estimated as the product of

U, surface area, temperature difference, and time. For buildings with temperature control a low U-value is normally desired; for heating equipment, a high U-value. The units for U-value are W/m^2K (Btu/ft^2 hr °F). Also see *Overall Heat Transfer Coefficient; R-Values*.

V

VACUUM DISTILLATION   A distillation process done at pressures below atmospheric with heat added to separate certain components. In petroleum refining, vacuum distillation is applied to the heavy distillates to further break them down. Also see *Heavy Distillate*.

VAPORIZATION   See Evaporation.

VAWT   Vertical axis wind turbine.

VEGOIL   Fuel made from vegetable oil.

VERTICAL AXIS WINDMILL   See Darrieus Rotor; Savonius Rotor.

VERY HIGH FREQUENCY (VHF)   Radio frequency of 30,000 to 300,000 Hz.

VERY LARGE CRUDE CARRIER (VLCC)   An oil tanker carrying over 150,000 long tons when loaded. Also see *Tankers*.

VISBREAKING   A process for reducing viscosity of a crude using a mild thermal process, less severe than coking, which produces a maximum of furnace oil.

VLCC   Very Large Crude Carrier. See Tankers.

VLSI   Very large scale integration.

VOID SPACE   The space between particles in a bulk of stored product usually expressed as percent of total volume.

Cereal grains	40-50%	Stone aggregates	25-55%
Coal	40-50%	Soybeans	30-40%
Sand, coarse, dry	20%	Wood	30-40%

Also see *Specific Gravity*.

VOLATILE SOLID A solid that can be converted to vapor by heating to a relatively low predetermined temperature. A complex material may contain some components that are volatile and others that are not. The two can be separated by heating as in the preparation of coke from coal.

More specifically, the volatile solid is the portion or quantity of solid which is volatilized by exposing to a temperature of 550° ± 50°C (1000° ± 90°F). The volatile solids is the difference between the total solid and the ash. One estimate is that 200-300 million tons of volatile solids are produced per year in the U.S.A. from 1.7 billion tons of municipal solids. *(Anderson, 1972.)*

VOLATILIZATION The process of converting a volatile solid to a vapor, usually by heating.

VOLTAGE CLASSIFICATION A system for classifying electric power according to the level of alternating current voltage.

Low voltage (LV)	1000 v or less
Medium voltage (MV)	1000 to 72,500 v
High voltage (HV)	72,500 to 242,000 v
Extra-high voltage (EHV)	242,000 to 800,000 v
Ultra-high voltage (UHV)	800,000 to 2,000,000 v

W

WASTE MATERIALS DRYING A wide variety of waste material can be made into useful by-products by drying. Garbage, waste from meat and fish processing, animal and plant wastes can be dried in a wide variety of dryers. The direct-fired horizontal rotary dryer is commonly used where an incoming temperature of 537.8°C (1000°F) is acceptable. Size reduction often follows drying.

WAO Wet air oxidation.

WAPA Western Area Power Administration. See Power Administrations.

WASTE OIL RECOVERY Oil used for engines and lubrication becomes contaminated with dirt, metal particles, and products of combustion. Approximately 5.3 billion liters (1.4 billion gallons) of used oil are produced each year. The lubricating oil does not wear out but incorporates undesirable contaminants. The used oil or waste oil can be processed or re-refined and used, saving approximately 1,470,000 liters (9,275 barrels) per day in the U.S.A. Approximately 9.5 liters (2.5 gal) of lubricating oil are obtained from 15.2 liters (4.0 gal) of used oil.

WATER A colorless, odorless, tasteless liquid, with formula H_2O, freezing point 0°C (32°F), and boiling point 100°C (212°F) at 1 atmosphere pressure (Table W-1). At 20°C (68°F) water has a viscosity of 1 centipoise; critical temperature, 374°C (705.2°F); critical pressure, 217.7 atmospheres; critical density, 0.4 gram/cm^3; specific heat, 1.0 at 15°C (59°F); dielectric constant, 81 at 18°C (64.4°F); thermal conductivity 0.32 Btu/ft^2 °F-hr/ft at 0°C (32°F), 0.36 at 37.8°C (100°F), and 0.39 at 93.3°C (200°F). A water molecule has a diameter of about 2.9 Å.

WATER GAS A low grade gaseous fuel (300 Btu/ft^3) used mainly by industry. Water gas is made by passing steam through hot coke. The

TABLE W-1
Water Properties

Density		lb/ft^3	kg/m^3
4°C	40°F	62.4	1000
20	68	62.3	998
Specific heat (heat capacity)		Btu/lb °F	kJ/kgK
20°C	68°F	1.0	4.19
Thermal conductivity		Btu/ft^2 °F hr/ft	W/mK
0°C	32°F	0.32	0.561
20	68	0.34	0.596
100	212	0.36	0.631
200	392	0.39	0.684
Vaporization, heat of		Btu/lb	kJ/kg
100°C	212°F at 1 atm.	970	2257.0
Viscosity, dynamic		centipoise	kg/ms
20°C	68°F	1.0	0.00101

Conversion factors: *1.0 Btu/ft hr °F = 1.7296 W/mK*
1.0 Btu/lb °F = 4184 J/kgK = 4.184 kJ/kgK
1.0 lb/ft^3 *= 16.0185* kg/m^3

steam partially oxidizes the coke and produces carbon monoxide and hydrogen according to the reaction:

$$C + H_2O \rightarrow CO + H_2$$

Water gas is composed of 40 to 45% carbon monoxide and 45 to 50% hydrogen with the remainder being mostly carbon dioxide and nitrogen. Because the water gas reaction is endothermic, it must be alternated with a combustion reaction during which air instead of steam is forced through the coke. Under conditions where this latter process yields only partial combustion, the product is producer gas, which may be mixed with the water gas to yield a product called semi-water gas. Also see *Carbureted Water Gas; Producer Gas.*

WATER VAPOR TRANSMISSION RATE (WVTR) The movement of moisture through products during heating or cooling or through a structural wall, expressed in weight of vapor transmitted per unit area per unit time, grains/ft^2-hr. Another term, permeance, perm is WVTR/Δp where Δp is the difference in inches Hg.

WATERWALL A boiler or heater in which the inner surfaces of the walls of the combustion chamber are made of vertical tubing carrying boiler feedwater upward as it is converted to steam.

WATERY ALCOHOL See Hydrous Alcohol.

WATT (W) A unit of power. One watt is one joule per second of work done, energy generated, energy used, etc. One joule/sec is 1/746 hp.

WATT-HOUR (Wh) A unit of energy or work. One watt hour is the energy or work generated, transmitted, or lost in one hour at a power level of one watt. The kilowatt hour (kWh), equal to 1000 watt hours, is the unit in which electrical energy use is commonly measured. One kilowatt hour is the electrical energy used by a 100 watt light bulb in 10 hours.

WAVE POWER The use of energy in water waves along coastal areas. A wave of 2-3 m (6.5-9.8 ft) with a period of 10 sec carries 4-9 kW per meter of wave width. The cost of wave energy is quite high at about 50¢/kWh. Also see *Ocean Thermal Energy Conversion (OTEC); Tidal Energy.*

WECS Wind Energy Conversion System.

WESCO Western Gasification Company project.

WESTERN INTERSTATE ENERGY BOARD (WIEB) An organization of 16 western states devoted to assisting the executive and legislative branches to develop all energy sources consistent with regional, social, economic, and environmental goals. Located at Stapleton Plaza, 3333 Quebec Street, Denver, CO 80207.

WESTERN OVERTHRUST BELT See Overthrust Belts.

WET AIR OXIDATION (WAO) A process in which the organic materials dissolved or suspended in water are converted to water and carbon dioxide. It is carried out at 204-315°C (400-600°F) and 600-1850 psi.

WET BASIS As used for moisture content. See Dry Basis; Moisture Content.

WET GAS; WET NATURAL GAS Natural gas, principally methane, which incorporates significant quantities, at least 2 gal of condensables per 1000 ft^3 by some standards, of hydrocarbon liquids such as ethane, propane, and butane. Also see *Associated Gas; Non-Associated Gas.*

WHOLE BODY RADIATION The average whole body radiation sources and levels for people in the U.S.A. are

	Mrem/y	Percent
Natural background	78	72
Medical	20	18
Fallout	8	7
Nuclear power industry	2	2
Consumer products	1	
	109	

WIEB Western Interstate Energy Board.

WIND ENERGY A term used both for the kinetic energy of the wind and the energy extracted from the wind by a wind turbine or other converter. The potential power from wind in the U.S.A. is shown in Figure W-1. The energy density of the wind, i.e., kinetic energy per unit volume of air, varies as the square of the wind speed, but the kinetic energy passing per unit time through a fixed wind turbine varies as the cube of the wind speed. An increase of a factor of two in wind speed raises the possible output of a wind turbine by a factor of eight. Thus, the potential for using wind energy at a given location varies greatly with wind speed. Ships propelled by the wind were common until the beginning of the 20th Century when they were replaced by coal-burning steamships. In the 11th Century wind was used to propel a bladed wheel which in turn rotated a shaft to operate a pump. Later applications of wind power included pumping water for drainage and irrigation, grinding and milling grain, and producing electricity. One estimate is that in 1850 in the U.S.A., windmills did 1.4×10^9 hp-hr of work in a year (equivalent to 11 million tonnes of coal). The World Meteorological Organization estimates the world's windpower at 20×10^9 kW. Also see *VAWT; Wind Turbine; Windmill; WTB.*

WIND TURBINE Also called wind generator, a device for converting the kinetic energy of the wind to electrical power. Wind turbines or generators were developed in the last 100 years. In comparison to windmills, wind turbines have fewer blades (2 or 3) and operate in winds of higher velocities (15-30 mph) with considerably higher top speeds, and have a higher efficiency (up to 45%). Direct current (DC) electricity is normally produced and used or stored in batteries. DC may be changed to AC by inverters. Wind turbines may also produce energy which can be stored by pumping water to a storage reservoir at a higher elevation, making hydrogen, or compressing air into a storage volume. One of the largest operating wind turbines, located in North Carolina, has 200 ft diameter blades, on a 150 ft tower. It generates up to 2000 kW of electricity and operates in 11 to 25 mph wind. Also see *Pumped Storage; Wind Energy; Windmill.*

WINDMILL A mechanical device consisting of a wheel on a horizontal shaft with blades driven by the wind to rotate a shaft to do work. Theoretically, 59% of the wind energy can be captured, and in practice 30-40% efficiency is obtained. Windmills, in contrast to wind turbines, have several blades, operate at medium to low wind speeds and are used for a variety of tasks. Two specialized windmills with the shaft in a vertical position are the Darrieus Rotor and the Savonius Rotor. Also see *Darrieus Rotor; Savonius Rotor; Wind Energy; Wind Turbine.*

WMO World Meteorological Organization.

WOBBE INDEX A relationship used to classify gases into similar

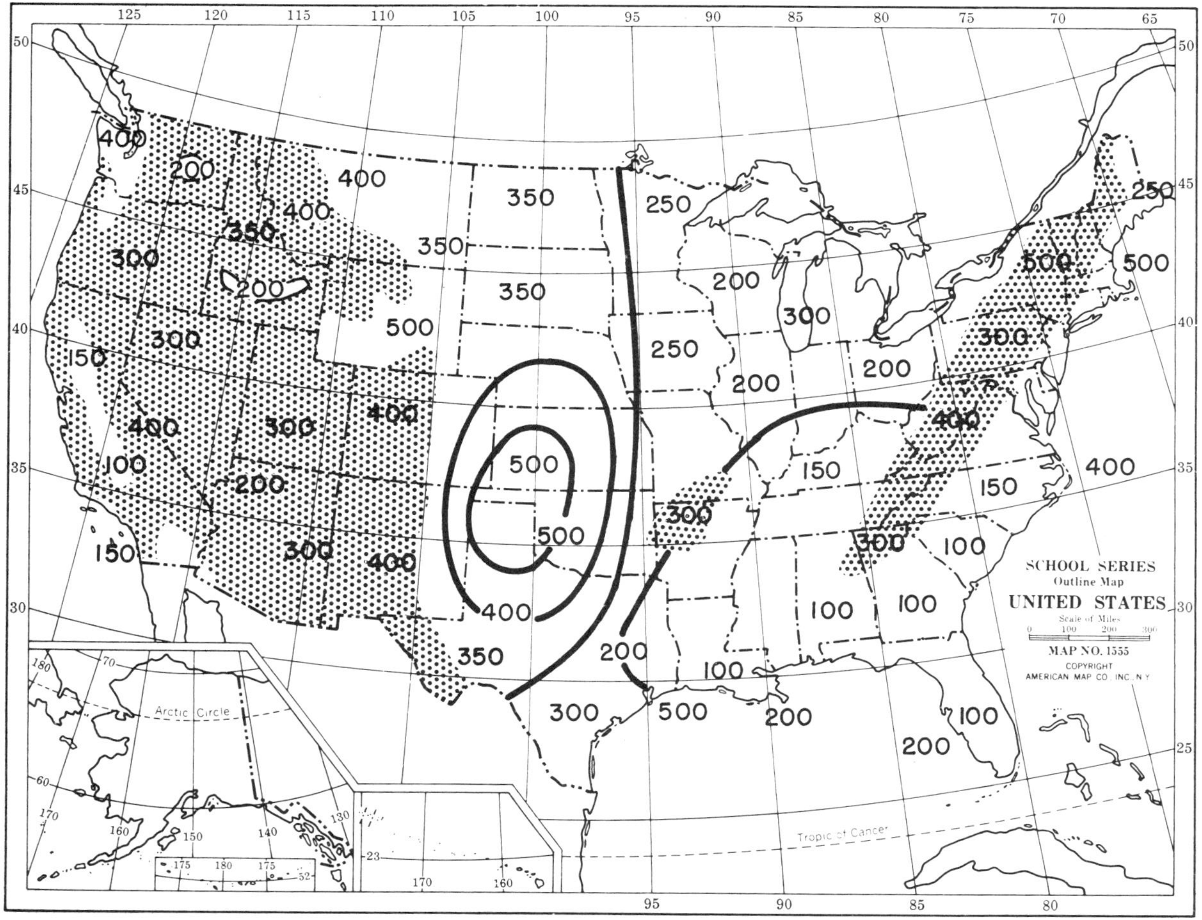

Fig. W-1 Wind power, mean annual values, W/m^2 *(Dickinson and Cheremisinoff, 1980).*

heating values, calculated by dividing the heating value by the square root of the specific gravity.

WOOD A lignocellulose material, naturally produced, consisting of 40-50% cellulose and 26% lignin. An empirical formula for wood is $C_6H_9O_4$ or $C_{0.32}H_{0.46}O_{0.22}$. Also see *Cellulose; Lignocellulose; Xylan.*

WOOD COMBUSTION The burning of the solid components of wood or bark which is preceded first by drying and second by distillation of other volatiles, the latter beginning at a temperature of about 275°C (527°F). The high heating (calorific) values of hardwoods are 18,850 to 20,446 kJ/kg (8100-8900 Btu/lb) and of softwoods 20,469 to 22,795 kJ/kg (8800-9800 Btu/lb).

WOOD ENERGY INSTITUTE P. O. Box 800, Camden, ME 04843.

WORK A measure of energy delivered or extracted from a mechanical system. Work done is equal to force along the direction of displacement times the size of the displacement and is measured in newton-meters (Nm) or foot-pounds, the newton-meter being 1 J. Also see *Energy, Joule; Potential Energy.*

WORT Solution of acid and sugars from a hydrolysis reaction on cellulose, or a mixture incorporating malt which becomes beer on fermentation.

WPPSS Washington Public Power Supply System.

WTC Whole tree chips.

WTG Wind turbine generator.

WVTR Water Vapor Transmission Rating.

X, Y, Z

X-RAY Short wavelength electromagnetic radiation emitted when electrons are accelerated or decelerated. X-rays of discrete wavelength are emitted when electrons make transitions between inner orbits in atoms. A continuous spectrum of x-rays is produced by a beam of high energy electrons impinging on a target material. The electrons in the beam are occasionally sharply deflected as they travel through matter. The large accelerations that result are accompanied by x-radiation (bremsstrahlung) which forms the continuous x-ray spectrum. The discrete x-ray wavelengths are characteristic of the atoms that produce them and can be used to identify materials. X-rays penetrate through matter and have been widely used to make visible on film the interior structures of the human body. Also see *Electromagnetic Spectrum*.

XYLAN A carbohydrate component in wood, associated with the cellulose. Xylan content is 15-20% in hardwood and 3-8% in softwood (*Kirk-Ottmer, 1970*). The general formula is $(C_5H_8O_4)_n$. Also see *Wood*.

XYLENE A heavy hydrocarbon of the benzene group used as a chemical feedstock. It is obtained from destructive distillation of coal and refining of oil and is also used as a blending agent in some gasolines.

YELLOW CAKE; YELLOWCAKE Usually described as U_3O_8, which is a yellow uranium oxide. However, the product of a uranium mill, also called yellow cake, is a mixture of oxides of uranium which may include $(NH_4)_2U_2O_7$ or $Na_2U_2O_7$, as well as U_3O_8. One ton of ore typically produces four pounds of yellow cake.

YIELD The quantity of a desired product produced from a given quantity of feedstock or by a given process. Thus, a bushel of cereal grain will yield 2.5 gal of ethanol. When product is expressed as a percent of the feedstock, percent yield or percent conversion is used.

A typical crude oil would yield:

Gasolines	50%
Kerosines, gas oils	34-35%
Fuel oils, lubricating oil, asphalt	10%
Used in refining and losses	6%

The yield of crude oil can be changed greatly by refining procedures depending on price structure and demand for various fractions. Cracking, catalytic cracking, thermal cracking, vacuum distillation, and hydrocracking are processes which are varied to change the mix of resulting products. Also see *Refinery*.

Z   A letter representing the atomic number of an element.

ZERO ENERGY BAND (ZEB)   A "dead band" of approximately 10°F where neither heating nor cooling occurs when controlling room temperature conditions. The use of a zero energy band in a control system is a means of reducing energy costs.

ZEROTH LAW OF THERMODYNAMICS   When two bodies are in thermal equilibrium with a third body, they are in thermal equilibrium with each other. Also see *Thermodynamics*.

ZINC-CHLORIDE ELECTRIC ENGINE   An electric engine unit using zinc and chlorine as active materials. Gulf and Western Industries has pioneered in the development of the system for electric vehicles, which uses commonly available materials. The chlorine is stored in the form of a hydrate of chlorine ($Cl_2 \cdot 8H_2O$). In comparison to the conventional lead-acid battery, the zinc-chloride system is claimed to be one-third as heavy and deliver four times the energy per unit weight.

ZIRCALOY   An alloy of zirconium widely used as a cladding for fuel elements to be used in nuclear reactors because of its low neutron absorption properties and high resistance to radiation damage and chemical corrosion. Zircaloy is 98% zirconium; 1.5% tin; 0.35% iron, chromium, and nickel; and 0.15% oxygen. Also see *Zirconium*.

ZIRCONIUM   A metallic element with atomic number 40 and atomic weight 91.2. Zirconium is widely used in the nuclear industry because of its low neutron absorption cross section and excellent resistance to radiation damage and corrosion. Also see *Zircaloy*.

ZYMURGY   A specialty of industrial chemical processes dealing with fermentation processes, such as making ethyl alcohol from carbohydrates.

REFERENCES

American Institute of Chemical Engineers. 1980. *Coal Processing Technology*. 6 volumes. New York, NY.

American Institute of Physics. 1975. *Efficient Use of Energy*. AIP Conference Proceedings No. 25. New York, NY.

American Society of Mechanical Engineers. 1957. *A Glossary of Terms in Nuclear Science and Technology*. New York, NY.

Anderssen, W. 1980. *Exergy*. Energy Technology (Sweden), No. 1, p. 15.

ASTM. 1978. *Test for the grindability of coal by the Hardgrove-Machine Method*. Am. Soc. Test. Mat. Standard D409. Philadelphia, PA.

Bailey, James E. 1978. *Energy Systems*. Marcel Dekker, Inc., New York, NY.

Baker, B. B., Jr., Deebel, W. R., and Geisenderger, G. D. 1966. *Glossary of Oceanographic Terms*. Special Publication 35, U. S. Naval Oceanographic Office, Washington, DC.

Baumeister, T., editor. 1967. *Mark's Standard Handbook for Mechanical Engineers*. McGraw-Hill Book Co., New York, NY.

Bentz, E. J. Jr., and Salmon, E. J. 1981. *Synthetic Fuels Technology Overview*. Ann Arbor Science, Ann Arbor, MI.

Besancow, R. M., editor. 1974. *The Encyclopedia of Physics*. Van Nostrand Reinhold Co., New York, NY.

Braunstein, H. M., Copenhaver, E. D., and Pfuderer, H. A. 1977. *Environmental health and control aspects of coal conversion: an information overview, v. 1.* ORNL/EIS-94. Oak Ridge National Laboratory, TN. April.

Cheremisinoff, P. N., and Moressi, A. C. 1976. *Energy from Solid Wastes.* Marcel Dekker, Inc., New York, NY.

Considine, D. M., editor. 1974. *Chemical and Process Technology Encyclopedia.* McGraw-Hill Book Co., New York, NY.

Considine, D. M., editor. 1976. *Scientific Encyclopedia.* Van Nostrand Reinhold Co., New York, NY.

Considine, D. M. 1977. *Energy Technology Handbook.* McGraw-Hill Book Co., New York, NY.

Corman, J. C. 1979. *Coal to electricity.* In 6th Energy Technology Conference & Exposition, Washington, DC.

Crabbe, David, and McBride, Richard. 1978. *The World Energy Book.* Nichols Publishing Co., New York, NY.

Crook, Leo. 1976. *Oil Terms.* International Publications Service, New York, NY.

Dickinson, W. C. and Cheremisinoff, P. N., editors. 1980. *Solar Energy Technology Handbook.* Marcel Dekker, Inc., New York, NY.

DOE. 1981. *Energy Technologies and the Environment.* Environmental Handbook (DOE/EP-0026), Washington, DC.

Dorf, R. C. 1978. *Energy, Resources, and Policy.* Addison Wesley Publishing Co., Reading, MA.

Duderstadt, J. J. 1979. *Nuclear Power.* Marcel Dekker, Inc., New York, NY.

Elliot, D. L. 1977. *Synthesis of national wind energy assessments.* Battelle Northwest Laboratories, BNW/Wind, Richland, WA.

Encyclopedia Britannica. 1961. Chicago, IL. 24 volumes.

Environment Information Center. 1974. *The Energy Directory.* Energy Reference Department, 124 E. 39th St., New York, NY.

Frauenfelder, H., and Henley, E. M. 1974. *Subatomic Physics.* Prentice Hall, Englewood Cliffs, NJ.

Gaynor, Frank, editor. 1950. *Concise Encyclopedia of Atomic Energy.* Philosophical Library, New York, NY.

Gloyna, E. F. 1980. *Long-range low-level waste management needs.* The University of Arizona, Symposium on Waste Management '80.

Tucson, AZ. March 10-14, 12 p.

Graves, Harvey. 1979. *Nuclear Fuel Management.* J. Wiley and Sons, New York, NY.

Hall, Carl W. 1979. *Dictionary of Drying.* Marcel Dekker, Inc., New York, NY.

Hall, C. W. 1981. *Biomass as an Alternative Fuel.* Government Institutes, Inc., Rockville, MD.

Hammond, A. L., Metz, W. D., and Maugh, T. H. 1973. *Energy and the Future.* American Association for the Advancement of Science, Washington, DC.

Hampel, C. A., and Hawley, G. D., editors. 1973. *The Encyclopedia of Chemistry.* Van Nostrand Reinhold Co., New York, NY.

Harper Encyclopedia of Science. 1967. Edited by James R. Newman. Harper and Row, New York, NY. 1379 p.

Harris, W. H., and Levey, J. S. 1975 (4th ed.). *The New Columbia Encyclopedia.* Columbia University Press, New York, NY.

Hatch, L. F., and Mazar, S. 1981. *From Hydrocarbons to Petrochemicals.* Gulf Publishing Co., Houston, TX.

Hogerton, J. F. 1963. *The Atomic Energy Deskbook.* Reinhold Publishing Co., New York, NY.

Holdren, J. P., and Herrera, P. 1971. *Energy: A Crisis in Power.* Sierra Club, San Francisco, CA.

Hokanson, A. E. 1977. *Chemicals from wood.* Pac. Chem. Engr. Congr. (PA-ChEC77) Aug. 28-31. Denver, CO. (Also Chem. Engr. Prog. 74 (1):67-71.)

Hottel, H. C., and Howard, J. B. 1971. *New Energy Technology—Some Facts and Assessments.* The MIT Press, Cambridge, MA.

Hunt, V. Daniel. 1979. *Energy Dictionary.* Van Nostrand Reinhold, New York, NY.

IAEA. 1976. *A short history of non-proliferation.* International Atomic Energy Agency, Vienna, Austria.

Jacobson, M., Nagy, J., Cooper, A. R., and Bau, F. J. 1961. *Explosibility of agricultural dusts.* U. S. Bureau of Mines RI 5753, U. S. Dept. of Interior, Washington, DC.

Jacobson, M., Nagy, J., and Cooper, A. R. 1962. *Explosibility of dusts used in the plastics industry.* U. S. Bureau of Mines RI

5971, U. S. Dept. of Interior, Washington, DC.

James, A. M. 1976. *A Dictionary of Thermodynamics*. Halsted Press, John Wiley & Sons, New York, NY.

Jenkins, Gilbert. 1977. *Oil Economists' Handbook*. Applied Science Publishers, Ltd., London.

Kirk, Raymond E., Othmer, D., et al. 1963-1970. *Encyclopedia of Chemical Technology*. Interscience Publishers, New York, NY. 22 volumes.

Kreiger, J. F., and Kreith, F. 1981. *Solar Energy Handbook*. McGraw-Hill Book Co., New York, NY.

Lapedes, D. N., editor. 1976. *McGraw-Hill Dictionary of the Life Sciences*. McGraw-Hill Book Co., New York, NY.

Lapedes, D. N. 1976. *Encyclopedia of Energy*. McGraw-Hill Book Co., New York, NY.

Lapedes, D. N., editor. 1978 (2nd ed.). *Dictionary of Scientific and Technical Terms*. McGraw-Hill Book Co., New York, NY.

Lindal, B. 1973. *Industrial and other applications of geothermal energy*. In C. H. Armstead, editor, UNESCO Press, New York, NY.

Loftness, R. L. 1978. *Energy Handbook*. Van Nostrand Reinhold Co., New York, NY.

McGraw-Hill Encyclopedia of Science and Technology. 1971. McGraw-Hill Book Co., New York, NY. 15 volumes.

Mandel, Siegfried. 1969. *Dictionary of Science*. Dell Publishing Co., New York, NY. 407 p.

Maramba, Felix D., Sr. 1978. *Biogas and Waste Recycling*. Liberty Mills, Metro Manila, Philippines.

Merriam, M. F. 1978. *Wind, waves, and tides*. Annual Review of Energy, v. 3, pp. 29-56.

Meyers, R. A. 1977. *Coal Desulfurization*. Marcel Dekker, Inc., New York, NY.

Miner, W. N. 1968. *Plutonium*. *Understanding the Atom Series*. U. S. Atomic Energy Commission, Washington, DC.

Mott-Smith, Morton. 1964. *The Concept of Energy Explained*. Dover Publications, Inc., New York, NY.

NAS., NRC. 1966. *Biological energy relationships and glossary of energy terms.* National Academy of Sciences Publ. 1411, Washington, DC.

National Bureau of Standards. 1975. *General Safety Standard for Installation Using Non-Medical X-Ray and Scaled Gamma Ray Sources.* NBS Handbook 114 (ANSI N543). U. S. Dept. of Commerce, Washington, DC.

Newman, J. R. 1967. *The Harper Encyclopedia of Science.* Harper & Row Publishers, New York, NY.

Nola, F. J. 1977. Power factor control system for AC induction motors. U. S. Patent 4052648.

Nunz, G. J. 1980. *Hot dry rock geothermal energy.* Mechanical Engineering 102(12):126-134.

Oppenheimer, E. J. 1980. *Natural Gas: The New Energy Leader.* Pen and Podium Productions, New York, NY.

Pelofsky, A. H., editor. 1977. *Heavy Oil Gasification.* Marcel Dekker, Inc., New York, NY.

Pelofsky, A. H., editor. 1979. *Coal Conversion Technology: Problems and Solutions.* I & EC Winter Symposium, Colorado Springs, CO. American Chemical Society, Washington, DC. 252 p.

Penner, S. S., and Icerman, L. 1974. *Energy.* (3 volumes.) Addison-Wesley Publishing Co., Reading, MA.

Pimentel, David. 1979. *Handbook of Energy Utilization in Agriculture.* CRC Press, Boca Raton, FL.

Prasad, K. N. 1974. *Radiation Biology.* Harper and Row Publishers, New York, NY.

Priest, Joseph. 1979. *Energy for a Technological Society.* Addison-Wesley Publishing Co., Reading, MA.

Public Affairs Clearinghouse. 1978. (2nd ed.). *Energy: A Guide to Organizations and Information Resources in the United States.* P. O. Box 10, Claremont, CA.

Schultz, V., and Whicker, F. W. 1980. *Nuclear fuel cycle, ionizing radiation, and effects on biota.* CRC Critical Reviews in Environmental Control 10(3):225-268.

Shreve, R. M., and Brink, J. A., Jr. 1977. *Chemical Process Industries.* McGraw-Hill Book Co., New York, NY.

Sullivan, Tom F. P. 1977. *Energy Reference Handbook.* Government Institutes, Inc., Rockville, MD.

Sussin, W. 1980. *Energy*. Scientific American 243(3):119-132.

Teller, Edward. 1979. *Energy from Heaven and Earth*. W. H. Freeman and Co., San Francisco, CA.

Thewlis, James. 1961-1964. *Encyclopedic Dictionary of Physics*. MacMillan, New York, NY. 9 volumes.

Thorpe, J. F., and Whitely, M. A. 1947-1954. *Thorpe's Dictionary of Applied Chemistry*. Longman, Green & Company, London and New York. 12 volumes.

Thrush, Paul W., et al., editor. 1968. *A Dictionary of Mining, Mineral, and Related Terms*. U. S. Dept. of Interior, Bureau of Mines. U. S. Government Printing Office, Washington, DC.

Tillman, D. A., Sarkanen, K. V., and Anderson, L. L. 1977. *Fuels and Energy from Renewable Resources*. Academic Press, New York, NY.

Todd, D. M., and Zeltmann, E. W. 1980. *Implementation of the fuel use acts*. Electric Forum 6(2):3-7.

Tubiana, M. 1980. *Respective risks of different energy sources*. IAE Agency Bull. 22(5/6):117.

Tvor, D. F., and Berry, R. W. 1980. *The Petroleum Dictionary*. Van Nostrand Reinhold Co., New York, NY.

U. S. Bureau of Mines. 1965. *Mineral facts and problems*. Bull. 630, U. S. Dept. of Interior, Washington, DC.

Van Oss, J. F. 1972. *Chemical Technology: An Encyclopedic Treatment*. Barnes and Noble, New York, NY. 6 volumes.

Weast, R. C., editor. 1979. *CRC Handbook of Chemistry and Physics*. CRC Press, West Palm Beach, FL.

Webster's New Collegiate Dictionary. 1975. G. & C. Merriam Co., Springfield, MA. 1536 p.

Whitehead, Harry. 1976. *An A-Z of Offshore Oil and Gas*. Kogan Page, London.

Whitehurst, D. D. 1980. *Coal Liquefaction*. Academic Press, New York, NY.

Wrage, K. E., and Goering, C. E. 1980. *Technical feasibility of diesohol*. Trans ASAE 23(6):1338-1343.